T0289544

Fundamentals of Modern Physics

Unveiling the mysteries

Online at: https://doi.org/10.1088/978-0-7503-6239-9

Fundamentals of Modern Physics

Unveiling the mysteries

Masatoshi Kajita
Graduates School of Arts and Science, The University of Tokyo, Tokyo, Japan

IOP Publishing, Bristol, UK

ISBN 978-0-7503-6239-9 (ebook)
ISBN 978-0-7503-6237-5 (print)
ISBN 978-0-7503-6240-5 (myPrint)
ISBN 978-0-7503-6238-2 (mobi)

DOI 10.1088/978-0-7503-6239-9

Version: 20231201

IOP ebooks

British Library Cataloguing-in-Publication Data: A catalogue record for this book is available from the British Library.

Published by IOP Publishing, wholly owned by The Institute of Physics, London

IOP Publishing, No.2 The Distillery, Glassfields, Avon Street, Bristol, BS2 0GR, UK

US Office: IOP Publishing, Inc., 190 North Independence Mall West, Suite 601, Philadelphia, PA 19106, USA

I learned the fundamental of laser spectroscopy from Professor T Shimizu when I was a graduate course student in the University of Tokyo. While I was with the National Institute of Information and Communications Technology (NICT), I learned a lot from discussions with Dr T Ido, Dr Y Yano, Dr S Nagano, Dr H Hachisu, Dr Y Li, and Dr M Kumagai all from NICT. After retirement from NICT, Professor T Aoki and Professor Y Torii gave me a change to continue research. I also greatly appreciate Professor F-L Hong (Yokohama National University) and Professor K Okada (Sophia Univiversity). I greatly appreciate all the people listed above.

Contents

Preface

Until the 19th century, only phenomena in the macroscopic world could be observed and they were mostly described with Newtonian mechanics and electromagnetism. At the end of the 19th century, surprising experimental results were discovered. One of these was the discovery of the constancy of the speed of light, which was made possible by improving the accuracy with which the speed of light was measured. The other was the discovery that the electron has negative electric charge and that its mass is much smaller than atoms. Atoms are not the smallest unit but are structures of electrons which have positive electric charge.

From the discovery of the constant speed of light, the theory of relativity was constructed. Most of the predictions from the theory of relativity have been experimentally confirmed. For example, an atomic clock with a measurement uncertainty below 10^{-17} was developed and the relativistic effects of time were confirmed (slower time in a moving frame or in a gravitational field). The gravitational wave was detected in 2015.

The first step for the establishment of quantum mechanics was the identification of the dual characteristic of light; as an electromagnetic wave and as a particle called a photon which has energy proportional to its frequency. There was a mystery also about the structure of atoms. It was clarified that atoms have a nucleus with a positive electric charge and electrons revolving around it. But the motion of the electron was not explained with any previous physics. Atoms were discovered to also absorb or emit light only with particular frequencies. These mysteries were solved when the wave characteristics of all particles was discovered. Particles bounded in a limited area can have only discrete energy because they can exist only with the wavelength resonant to the size of the area. Since the development of the laser, spectroscopic experiments have been performed and the energy structure of many atoms and molecules have been clarified. After the development of laser cooling technology, significant quantum effects have been observed with atoms; atomic interferometry, quantum effects with atomic collision, Bose–Einstein condensation, realization of entangled states (correlation between states of multi-particles), and Schrödinger's cat phenomenon (single particles can exist at two different positions simultaneously).

But there are still mysteries which have not been solved. Since the establishment of relativistic quantum mechanics, the existence of antiparticles was predicted and experimentally confirmed. Pair production and pair annihilation were also observed. The characteristics of antiparticles looks like the charge conjugation + mirror image of particles (CP-symmetry). But if the CP-symmetry holds perfectly, the abundance of particle and antiparticles should be exactly equal. The violation of CP-symmetry has been discovered. The Kobayashi–Maskawa theory derived the possible violation of the CP-symmetry assuming the existence of the three generation quarks, which were discovered afterwards. But there are still some mysteries about the violation of the CP-symmetry which have not been solved.

The shape of the Universe was believed to be immutable until the expansion of the Universe was discovered. It was clarified that the universe was born from the explosion 13.8 billion years ago (called the Big Bang). The acceleration of the expansion speed was discovered, which created another mystery which was named 'dark energy'. On the other hand, the gravitational interaction in the Universe was understood to be much stronger than that predicted from the mass of all visible matter. Therefore, there must be some material which cannot be observed with light, called 'dark matter'. The identities of dark energy and dark matter are important mysteries in modern physics.

This book introduces these mysteries of modern physics targeting mainly graduate course students. In chapter 1, the fundamental theory of relativity, quantum mechanics, and relativistic quantum mechanics are simply introduced. This knowledge is required to understand the mystery of modern physics. Chapter 2 introduces the experimental procedures of modern physics; precision measurement of atomic or molecular transition frequencies and confirmation of relativistic or quantum effects. The role of the development of the laser which had a significant effect on experimental physics is discussed. Chapter 3 introduces the fundamentals of the elementary particle and the mystery of the non-equal abundance ratio between particles and antiparticles. Chapter 4 introduces the mystery of astrophysics with dark energy and dark matter.

In this book, equations have been minimized so that the concepts are simple for readers to understand.

Acknowledgements

The research activity of the author has been supported by a Grant-in-Aid for Scientific Research (B) (Grant No. JP 17H02881 and JP20H01920), and a Grant-in-Aid for Scientific Research (C) (Grant Nos. JP 17K06483 and 16K05500) from the Japan Society for the Promotion of Science (JSPS). The author is highly appreciative of discussions with Y Yano, T Ido, N Ohtsubo, A Shinjo-Kihara, H Hachisu, S Nagano, M Kumagai, S Hayashi, N Sekine, and M Hara, all from the NICT, Japan, as well as K Okada (Sophia U.), T Aoki and Y Torii (the University of Tokyo), and N Kimura (RIKEN). The author is grateful to K Kameta and J Navas (IOP, UK) for the opportunity to write this book.

Author biography

Masatoshi Kajita

Born and raised in Nagoya, Japan, Dr Kajita graduated from the Department of Applied Physics, the University of Tokyo in 1981 and obtained his PhD from the Department of Physics, the University of Tokyo in 1986. After working at the Institute for Molecular Science, he joined the Communications Research Laboratory (CRL) in 1989. In 2004, the CRL was renamed the National Institute of Information and Communications Technology (NICT). In 2009, he was guest professor at the Provence Universite, Marseille, France. In 2023, he retired from NICT and became the guest researcher in the University of Tokyo.

Chapter 1

Introduction to the fundamental parts of modern physics

Since the beginning of the 20th century, there has been a revolution in physics due to the development of the theory of relativity and quantum mechanics (modern physics). There are still some mysteries which have not been solved in modern physics; these subjects are introduced in this book. To understand the current situation in physics research, it is necessary to understand the fundamentals of modern physics. In section 1.1, the history of the development of physics up to Newtonian mechanics is introduced. Section 1.2 introduces the identity of light, which provided the opportunity to develop modern physics. Sections 1.3 and 1.4 introduce the fundamentals of the special and general theories of relativity, respectively. Section 1.5 introduces the fundamentals of quantum mechanics. Relativistic quantum mechanics is introduced in section 1.6.

1.1 History of physics up to the establishment of Newtonian mechanics

Physics is a research field which searches for laws with which we can predict phenomena in future. The laws are established based on our previous experiences. Some discrepancies with previous laws have been discovered when new phenomena were observed; this meant that new physics was required to explain the new phenomena.

How were the new phenomena discovered? The most important thing is the reduction of measurement uncertainties. With high measurement uncertainties, we establish physical laws very roughly. We tend to give the simplest interpretations, but these might not be valid after new phenomena are discovered when measurement uncertainty is reduced. The expansion of the observation range also had an important role in establishing new physics, because physical laws established in a

doi:10.1088/978-0-7503-6239-9ch1

limited area sometimes had discrepancies with the observations in the new area. Newtonian mechanics was established with the procedure as shown below.

Research into natural philosophy started in ancient Greece. Aristotle argued that the Earth was spherical for the simple reason that going south, constellations along the southern horizon are seen at higher positions, and that we observe a circular shadow of the Earth at lunar eclipse. However, the Earth was believed to be the center of the Universe, about which the Sun and the stars were believed to revolve. The most famous theory is that of Ptolemy. At that time, the uncertainty in the telling of the time of day (using a sundial or water clock) was of the order of 1 h, and no discrepancy between the observed positions of stars and the Ptolemaic theory was discernible [1]. After the accuracy of time keeping was improved, the discrepancies between Ptolemaic theory and observations were realized. In 1510, Copernicus published his idea that the Earth revolved around the Sun. However, Copernican theory could not describe the observation results in detail. This is because he assumed that the planets, including the Earth, had circular orbits.

In 1581, Galilei discovered the periodicity of the pendulum's swing, which made it possible to reduce the uncertainty of time to the level of 10 min per day (later improved to 1 min per day). In 1609, Galilei observed the four satellites of Jupiter using a handmade telescope. This was the first observation of phenomena outside of the Earth; this provided a step towards establishing the Copernican theory. This was finally established between 1609 and 1619 when Kepler published the following three laws of planetary motion:

1. The orbit of a planet is an ellipse with the Sun at one of the two foci.
2. A line segment joining a planet and the Sun sweeps out equal areas during equal time intervals.
3. The square of the orbital period of a planet is proportional to the cube of the semi-major axis of orbit.

Newton derived the three laws of motion (first published in 1687).

1. An object either remains at rest or continues to move at a constant velocity, unless acted on by a force.
2. The sum of the forces $\vec{F}$ on an object is equal to the mass m of that object multiplied by the acceleration $\vec{a}$ of the object,

$$\vec{F} = m\vec{a}. \tag{1.1.1}$$

3. When one body exerts a force on a second body, the second body simultaneously exerts a force equal in magnitude and opposite in direction on the first body.

In addition, he presented the law of universal gravitation (published in 1687) showing the gravitational attractive force between two bodies F_G takes the form

$$F_G = GMm/r^2 \tag{1.1.2}$$

where G is the gravitational constant, M and m are the masses of two bodies, and r is the distance between the two bodies. Newtonian mechanics was established roughly in the same period as the invention of the pendulum clock, which was the most accurate clock up to the 1930s. The establishment of Newtonian mechanics was correlated with the accuracy improvement of clocks, which made it possible to observe of the phenomena outside of the earth in detail.

In the 18–19th centuries, the fundamentals of electromagnetism were established. With the improvement of experimental technology, it became possible to observe the phenomena also in the microscopic world. It was realized that the phenomena in the microscopic world cannot be described by Newtonian mechanics and quantum mechanics was established. The speed of light was realized to be constant to all observers not in motion, and the theory of relativity was established. There are also some mysteries, which have still not yet been solved. This book introduces a few examples of mysteries in modern physics after introducing the fundamentals of the theory of relativity and quantum mechanics in this chapter.

1.2 Mystery of light

Light has a very important role in our lives, as we cannot see anything without it. However, it was not clearly understood until the 19th century. In the 17th century, 'particle theory' and 'wave theory' were proposed. Wave theory seemed to be more advantageous from the observation of diffraction. The observation of interference also made wave theory advantageous, but it was not reproduceable. However, nobody could explain what kind of oscillation propagates in what kind of medium. Therefore, wave theory was not decisively advantageous.

The understanding of the identity of light caused the appearance modern physics: the theory of relativity and quantum mechanics. As shown below, the speed of light in a vacuum is constant. And light has dual characteristics both as a wave and a particle. There was still the mystery of why light has such special characteristics. But the theory of relativity indicates that all particles without mass must move at the speed of light. Quantum mechanics indicates that all particles have dual characteristics.

1.2.1 Speed of light

While considering the phenomena observed until the middle of the 17th century, there were no discrepancies by considering the speed of light to be infinite. The first phenomenon which indicates a finite speed of light was the fluctuation of the period taken by Io, one of Jupiter's satellites, to complete an orbit. The reduction of measurement uncertainty of the period made it possible to discover this phenomenon. In 1676, Rømer considered a finite propagation time for light between Jupiter and the Earth [2]. The distance between the Earth and Jupiter varies with their orbital motions around the Sun. The propagation time of light from Jupiter also varies while Io revolves around Jupiter as shown in figure 1.1. With this idea, the speed of light was estimated to be 2.2×10^8 m s^{-1}, which is 26% lower than the present value.

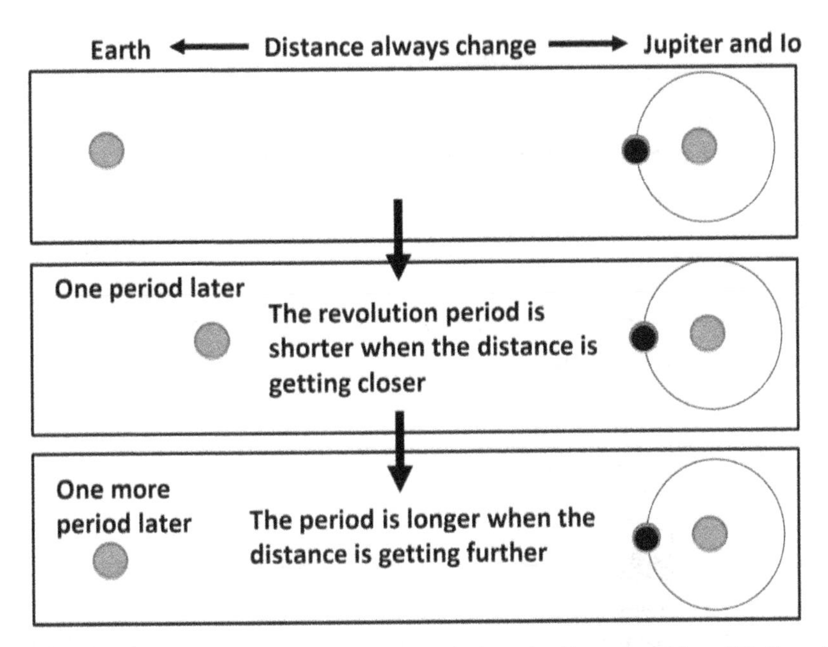

Figure 1.1. The reason Io's orbital period was observed to be irregular. Reproduced from [31].

In 1727, Bradley estimated the speed of light from the changes of the direction of light propagation from the Draco γ star induced by the Earth's orbital motion. The estimated speed of light was 3.01×10^8 m s^{-1}, which is only a few percentage points different from the present value [3].

The first successful terrestrial measurement of the speed of light was performed by Fizeau in 1849 [4]. Light passing through a gear was reflected by a mirror 9 km away (figure 1.2). The speed of light was determined from the velocity of rotation of the gear for when the reflected light is blocked and found to be 3.13×10^8 m s^{-1}. In 1862, Foucault measured the speed of light using the same principle as that used in Fizeau's experiment except he used a rotating mirror instead of a gear. The measurement gave 2.98×10^8 m s^{-1}, which is 0.6% different from the present value [5].

The measurement of the speed of light made it possible to search for the identity of light, as shown below.

1.2.2 Light as an electromagnetic wave

The laws of electromagnetism regarding the electric field $\vec{E}$ and the magnetic field $\vec{B}$ are summarized by Maxwell's equations shown below (in SI units) [6].

$\nabla \cdot \vec{E} = \frac{\rho}{\varepsilon}$ (ρ: the electric charge density, ϵ: the permittivity)

$$\frac{\partial E_x}{\partial x} + \frac{\partial E_y}{\partial y} + \frac{\partial E_z}{\partial z} = \frac{\rho}{\varepsilon} \text{ (derived from Coulomb's law)} \tag{1.2.1}$$

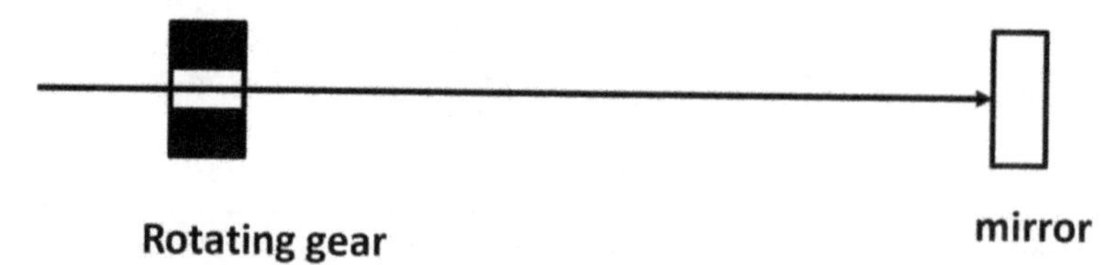

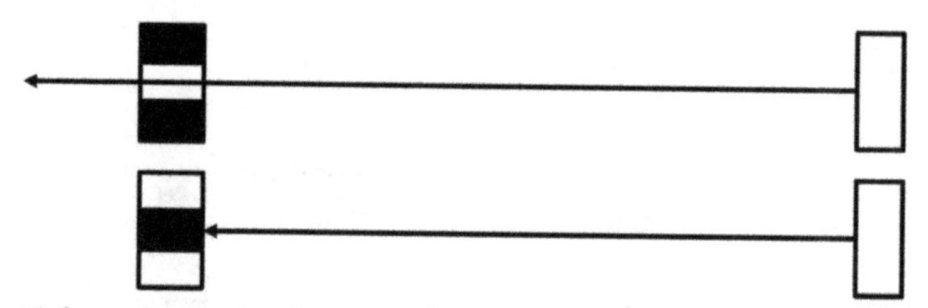

Figure 1.2. Principle involved in the measurement of the speed of light by Fizeau. Reproduced from [31].

$$\nabla \times \vec{E} = -\frac{\partial \vec{B}}{\partial t}$$

$$\frac{\partial E_z}{\partial y} - \frac{\partial E_y}{\partial z} = -\frac{\partial B_x}{\partial t}, \frac{\partial E_x}{\partial z} - \frac{\partial E_z}{\partial x} = -\frac{\partial B_y}{\partial t}, \frac{\partial E_y}{\partial x} - \frac{\partial E_x}{\partial y} = -\frac{\partial B_z}{\partial t} \text{ (derived from Faraday's law of induction)} \quad (1.2.2)$$

$$\nabla \cdot \vec{B} = 0$$

$$\frac{\partial B_x}{\partial x} + \frac{\partial B_y}{\partial y} + \frac{\partial B_z}{\partial z} = 0 \text{ (there is no magnetic charge)} \quad (1.2.3)$$

$$\nabla \times \vec{B} = \mu\left[\vec{j} + \varepsilon\frac{\partial \vec{E}}{\partial t}\right] \text{ (}\vec{j}\text{: the electric current density, }\mu\text{: the permeability)}$$

$$\frac{\partial B_z}{\partial y} - \frac{\partial B_y}{\partial z} = \mu\left[j_x + \varepsilon\frac{\partial E_x}{\partial t}\right], \frac{\partial B_x}{\partial z} - \frac{\partial B_z}{\partial x} = \mu\left[j_y + \varepsilon\frac{\partial E_y}{\partial t}\right], \frac{\partial B_y}{\partial x} - \frac{\partial B_x}{\partial y} = \mu\left[j_z + \varepsilon\frac{\partial E_z}{\partial t}\right] \quad (1.2.4)$$

(derived from Ampere's circuital law, including around the condenser)

Maxwell's equations are just the change of descriptions of laws, which were already known. However, this equation made a revolution in electromagnetism because the electromagnetic wave is derived as follows. Taking the AC electric field direction in the x-direction, equation (1.2.2) gives

$$\frac{\partial E_x}{\partial z} = -\frac{\partial B_y}{\partial t}, \frac{\partial E_x}{\partial y} = \frac{\partial B_z}{\partial t}, \quad (1.2.5)$$

and equation (1.2.4) with $\vec{j} = 0$ is expressed as

$$\frac{\partial B_z}{\partial y} - \frac{\partial B_y}{\partial z} = -\varepsilon\mu\frac{\partial E_x}{\partial t}. \tag{1.2.6}$$

The AC electric field in the x-direction induces an AC magnetic field in the y- or z-direction; here we take the AC magnetic field in the y-direction. Thus, we have

$$\begin{aligned} &\frac{\partial E_x}{\partial z} = -\frac{\partial B_y}{\partial t} \rightarrow \frac{\partial^2 E_x}{\partial z^2} = -\frac{\partial^2 B_y}{\partial z \partial t} \\ &\frac{\partial B_y}{\partial z} = -\varepsilon\mu\frac{\partial E_z}{\partial t} \rightarrow \frac{\partial^2 B_y}{\partial z \partial t} = -\varepsilon\mu\frac{\partial^2 E_x}{\partial t^2} \\ &\frac{\partial^2 E_x}{\partial z^2} = \varepsilon\mu\frac{\partial^2 E_x}{\partial t^2}. \end{aligned} \tag{1.2.7}$$

To solve equation (1.2.7), E_x is assumed to be the product of the functions Z_E and T_E, which depend only on z and t, respectively. Then we have:

$$\begin{aligned} &T_E\frac{\partial^2 Z_E}{\partial z^2} = \varepsilon\mu Z_E\frac{\partial^2 T_E}{\partial t^2}, \quad \frac{1}{\varepsilon\mu}\frac{1}{Z_E}\frac{\partial^2 Z_E}{\partial z^2} = \frac{1}{T_E}\frac{\partial^2 T_E}{\partial t^2} = C \\ &T_E = a_+ e^{\sqrt{C}t} + a_- e^{-\sqrt{C}t} \quad Z_E = b_+ e^{\sqrt{\varepsilon\mu C}z} + b_- e^{-\sqrt{\varepsilon\mu C}t}. \end{aligned} \tag{1.2.8}$$

When $C < 0 (C = -\Omega^2)$,

$$\begin{aligned} E_x &= c_1 e^{i\Omega(t+\sqrt{\varepsilon\mu}z)} + c_2 e^{i\Omega(t-\sqrt{\varepsilon\mu}z)} + c_3 e^{-i\Omega(t+\sqrt{\varepsilon\mu}z)} + c_4 e^{-i\Omega(t-\sqrt{\varepsilon\mu}z)} \\ &= E_0 \sin\left[\Omega\left(t \pm \sqrt{\varepsilon\mu}z\right) + \varphi_0\right]. \end{aligned} \tag{1.2.9}$$

Equation (1.2.9) shows that E_x is a waveform that propagates in the z-direction with a velocity of $c_m = \mp\frac{1}{\sqrt{\varepsilon\mu}}$. For B_y, we have:

$$\begin{aligned} &\frac{\partial E_x}{\partial z} = -\frac{\partial B_y}{\partial t}, \frac{\partial B_y}{\partial z} = -\varepsilon\mu\frac{\partial E_z}{\partial t} \rightarrow B_y \\ &= B_0 \sin\left[\Omega\left(t \pm \sqrt{\varepsilon\mu}z\right) + \varphi_0\right] B_0 = -\sqrt{\varepsilon\mu}E_0. \end{aligned} \tag{1.2.10}$$

The estimated propagation speed $c_m = \frac{1}{\sqrt{\varepsilon\mu}}$ corresponds to the speed of light, as measured by Fizeau and Foucault [4, 5]. Therefore, light is proven to be an electromagnetic wave. A change in the electric field induces a change in the magnetic field, and vice versa. The change in the electric and magnetic fields propagates as waves. In a propagating wave, the phases of the oscillation of the electric and magnetic fields are equal. As shown above, the directions of electric field, magnetic field, and the propagation direction are orthogonal each other. In the same propagation direction (z), two kinds of polarization are possible: (E_x,B_y) or (E_y,B_x). The interference between two light waves is caused only between same polarization, therefore the interference is not always observed. The linear polarization indicates that the direction of the electric (or magnetic) field is constant. The circle polarization indicates that the direction of the electric (or magnetic) field rotates in the right (σ^+) or left (σ^-) direction in the propagation direction.

There was a mystery with the speed of light, given by $c_m = \frac{1}{\sqrt{\varepsilon\mu}}$. The observed velocity depends on the motion of observers. We assume that a velocity of V is observed by A. When B is moving with the velocity of v against A, B would observe the velocity of V–v. When the light propagates in a medium (given by a material), c_m is the speed against the medium. For propagation in a vacuum, what is the medium that serves as the standard for the speed of light? The values of ε and μ in a vacuum space (ε_0 and μ_0) are universal constants, and the speed of light in a vacuum $c = \frac{1}{\sqrt{\varepsilon_0\mu_0}}$ derived from Maxwell's equation does not depend on the motion of the observer. To solve this discrepancy, the speed of light in a vacuum was defined as the speed against the medium called ether. If we move against the ether, the speed of light is expected to shift. However, this effect was estimated to be too small to be detected because we only experience velocities much slower than the speed of light.

Michelson and Morley expected that this effect could be detected comparing the round propagation times in the parallel and perpendicular directions to the Earth's orbital motion ($v_{\rm orb}$) [7]. The round propagation time in the parallel and perpendicular directions are given by $\tau_{\|} = 2L_{\|}/(c^2 - v_{\rm orb}^2)$ and $\tau_{\perp} = 2L_{\perp}/\sqrt{c^2 - v_{\rm orb}^2}$, respectively ($L_{\|,\perp}$: distance to the mirror to the parallel and perpendicular direction). With the orbital motion of the Earth (change of direction of motion), the variation in $\Delta\tau = (\tau_{\|} - \tau_{\perp})$ was expected to be observed by the variation in the interference signal between the lights propagated in both directions. The interference signal indicates the phase difference between two optical paths $\Delta\varphi_{ph} = 2\pi\nu(\Delta\tau)$, where ν is the frequency of the light (with this experiment, Na lamp). The observation of the interference signal is advantageous to observe the slight variation of $\Delta\tau$, because it focuses after the decimal point of $\Delta\varphi_{ph}/2\pi=\nu(\Delta\tau)$. The variation in the speed of light according to the Earth's orbital motion (speed of $10^{-4}c$) was expected to be detected with this system. However, the speed of light was measured to be constant within a fractional uncertainty of 10^{-8}. Using a laser source, the constancy of the speed of light has been confirmed with an uncertainty of 10^{-15} [8]. The constancy of the speed of light remained a mystery until the establishment of the theory of relativity.

1.2.3 Characteristic of light as a particle

The identity of light was clarified to be an electromagnetic wave. However, some phenomena could not be explained by considering light only as a wave. When a material is heated, emission of light is observed: this irradiation is called 'blackbody radiation'. White thermoelectric balls are light sources using blackbody radiation. The spectrum distribution of the blackbody radiation (shown in figure 1.3) has the following characteristics [9].

(1) The intensity is maximum at a certain frequency, which is proportional to the thermodynamic temperature T (Celsius temperature + 273.15 K) and the total intensity is proportional to T^4. The blackbody radiation with room

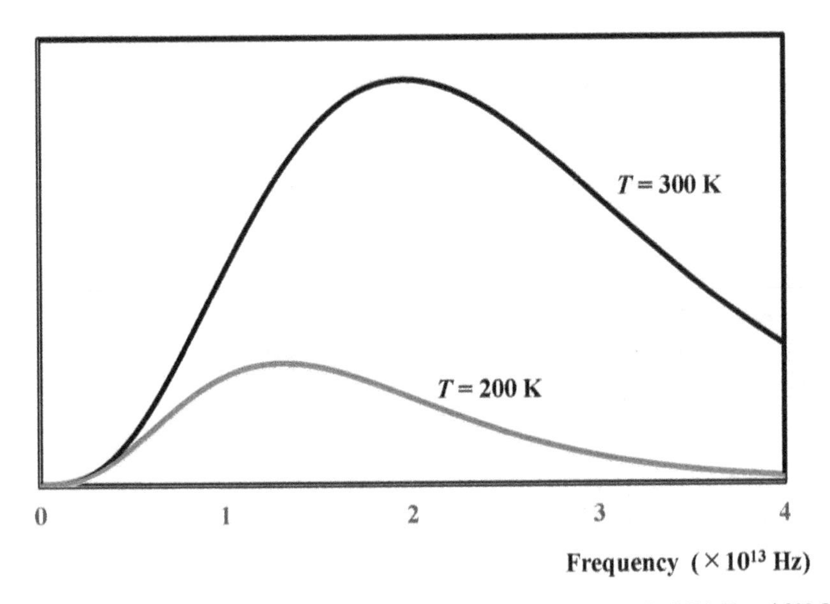

Figure 1.3. Spectrum distribution with the thermodynamic temperature T of 300 K and 200 K.

temperature is not visible, because it distributes mostly in the infrared region (visible using IR viewer).

(2) The energy density at the low frequency region is given by $\frac{8\pi\nu^2}{c^3}k_BT$, where ν is the frequency, and k_B is the Boltzmann constant (1.38×10^{-23} J K^{-1}). This characteristic is explained by the pure wave characteristic of the radiation.

(3) With the high frequency region, the distribution is proportional to $\nu^3 \exp\left[-\frac{h\nu}{k_BT}\right]$. Here, h is the Planck constant (6.6×10^{-34} J Hz^{-1}). This characteristic cannot be explained considering light only as a wave.

These results indicate that the properties of light cannot be described as a pure wave characteristic with high frequency. Planck derived a general formula of the spectrum distribution as follows based on the assumption that the light energy can only assume $n_\alpha h\nu$ with the probability proportional to $\exp\left(-\frac{n_\alpha h\nu}{kBT}\right)$, where $n_\alpha(\geqslant 0)$ is integer.

$$P_{BBR} = \frac{8\pi h\nu^3}{c^3}\frac{1}{\exp\left[\frac{h\nu}{k_BT}\right] - 1}, \tag{1.2.11}$$

which is in good agreement with the experimental result in the whole frequency area. The detailed derivation of equation (1.2.11) is shown in appendix A. The previous results (2) and (3) were interpreted as the approximations of equation (1.2.11) with

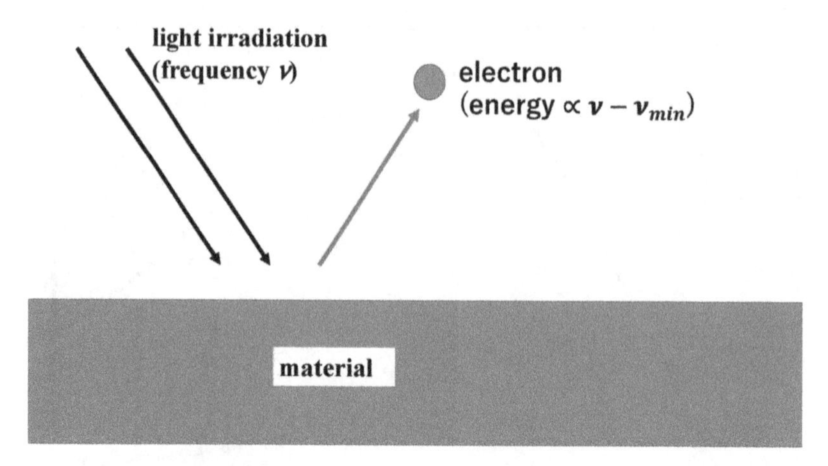

Figure 1.4. The concept of the photoelectronic effect. When light with a frequency ν higher than the minimum threshold frequency ν_{min} is irradiated, electrons are emitted. The energy of emitted electron is proportional to $\nu - \nu_{\text{min}}$.

the limits of low and high frequencies. However, the validity of Planck's assumption was not confirmed at that time.

In addition, the emission of electrons from a material was observed when light is irradiated on the surface (called the photoelectronic effect shown in figure 1.4). The experimental results for the photoelectric effect show that the energy of the emitted electrons is independent of the intensity of the light, although the number of emitted electrons is proportional to the intensity. Moreover, emission does not occur when the frequency of light ν is lower than a minimum threshold value ν_{min} and the energy of the emitted electron is proportional to the $\nu - \nu_{\text{min}}$. Einstein proposed a new concept of wave–particle duality; light has the characteristics of both waves and particles. The energy of each particle (called a photon) is $E = h\nu$ and the momentum is $\vec{p} = h\vec{k}$ $(|\vec{k}| = 1/\lambda)$, where $\vec{k}$ is the wavenumber vector and λ is the wavelength. Planck's assumption was also explained by the particle–wave duality of light: light energy is product of the photon number n_α and the energy of each photon $h\nu$. This duality was a special characteristic of light until the concept of matter waves was proposed.

1.3 Theory of special relativity

Many scientists tried to explain the constancy of speed of light c, for example by 'compression of ether by motion'. While then Einstein established the special theory of relativity [10] based on the constancy of the speed of light in a vacuum. We consider the coordinates (x,y,z,t) and (x',y',z',t'); (x',y',z',t') are coordinates of an observer moving with a velocity of v on the x-direction, as shown in figure 1.5. For classical mechanics, the relationship between both coordinates is given by

$$x' = x - vt,\ y' = y,\ z' = z,\ t' = t, \tag{1.3.1}$$

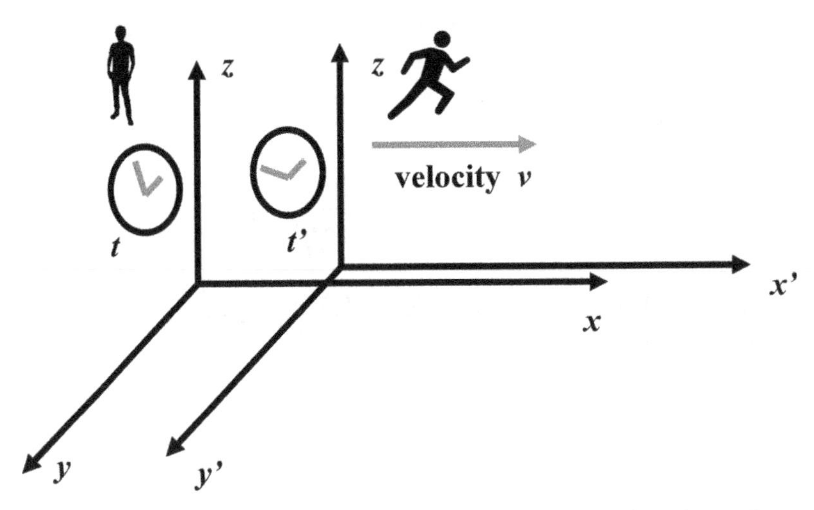

Figure 1.5. Coordinates (x,y,z,t) and (x',y',z',t'), where (x',y',z',t') is the coordinate for an observer moving with the velocity of v on the x-direction.

and the speed of light changes with $c' = c - v$. Another formula is required for the translation between different coordinates while maintaining a constant speed of light. The fundamentals of the theory of special relativity is the identity between $(ct')^2 = x'^2 + y'^2 + z'^2$ and $(ct)^2 = x^2 + y^2 + z^2$. For simplicity, we assume $y' = y$ and $z' = z$ and consider the identity between $ct' = x'$ and $ct = x$ using the matrix:

$$\begin{pmatrix} x' \\ ct' \end{pmatrix} = \begin{pmatrix} p & q \\ r & s \end{pmatrix}\begin{pmatrix} x \\ ct \end{pmatrix}. \tag{1.3.2}$$

with the requirement

$ct' = x' \rightarrow rx + sct = px + qct, \quad (s - q)ct = (p - r)x \rightarrow s - q = p - r$

$q = -p\frac{v}{c}, \;\; s = p$ (with $v \ll c$, equation (1.3.1)) $\rightarrow r = -\frac{v}{c}p$

$$\begin{pmatrix} x' \\ ct' \end{pmatrix} = p\begin{pmatrix} 1 & -\frac{v}{c} \\ -\frac{v}{c} & 1 \end{pmatrix}\begin{pmatrix} x \\ ct \end{pmatrix} \tag{1.3.3}$$

For $v \rightarrow -v$, the matrix is transformed to an inverse matrix.

$$p^2\begin{pmatrix} 1 & \frac{v}{c} \\ \frac{v}{c} & 1 \end{pmatrix}\begin{pmatrix} 1 & -\frac{v}{c} \\ -\frac{v}{c} & 1 \end{pmatrix} = \begin{pmatrix} 1 & 0 \\ 0 & 1 \end{pmatrix} p^2 = \frac{1}{1 - \left(\frac{v}{c}\right)^2}$$

$$\begin{pmatrix} x' \\ ct' \end{pmatrix} = \frac{1}{\sqrt{1 - \left(\frac{v}{c}\right)^2}}\begin{pmatrix} 1 & -\frac{v}{c} \\ -\frac{v}{c} & 1 \end{pmatrix}\begin{pmatrix} x \\ ct \end{pmatrix}. \tag{1.3.4}$$

The transformation of the four-dimensional vector (x, y, z, ct) is given by:

$$\begin{pmatrix} x' \\ y' \\ z' \\ ct' \end{pmatrix} = \begin{pmatrix} \frac{1}{\sqrt{1-(v/c)^2}} & 0 & 0 & \frac{-(v/c)}{\sqrt{1-(v/c)^2}} \\ 0 & 1 & 0 & 0 \\ 0 & 0 & 1 & 0 \\ \frac{-(v/c)}{\sqrt{1-(v/c)^2}} & 0 & 0 & \frac{1}{\sqrt{1-(v/c)^2}} \end{pmatrix} \begin{pmatrix} x \\ y \\ z \\ ct \end{pmatrix}, \tag{1.3.5}$$

which is called the 'Lorenz transformation'. What happens to the length and the time interval in a moving coordinate system?

$$\begin{aligned} x'_1 - x'_2 &= \frac{1}{\sqrt{1-\left(\frac{v}{c}\right)^2}}[(x_1 - vt_1) - (x_2 - vt_2)] \\ ct'_1 - ct'_2 &= \frac{1}{\sqrt{1-\left(\frac{v}{c}\right)^2}}\left[\left(-\frac{v}{c}x_1 + ct_1\right) - \left(-\frac{v}{c}x_2 + ct_2\right)\right] \end{aligned} \tag{1.3.6}$$

The length in the moving coordinate system should be considered with $t'_1 = t'_2$ and

$$t_1 - t_2 = \frac{v}{c^2}(x_1 - x_2)$$

$$x'_1 - x'_2 = \sqrt{1-\left(\frac{v}{c}\right)^2}(x_1 - x_2) \tag{1.3.7}$$

is derived. The time interval for $x'_1 = x'_2$ is

$$t'_1 - t'_2 = \sqrt{1-\left(\frac{v}{c}\right)^2}(t_1 - t_2) \tag{1.3.8}$$

Figure 1.6 shows the idea that time goes slower in a moving frame, as the propagation time in the direction perpendicular to the motion is longer because the absolute value of speed of light is c and its component in the perpendicular direction is $\sqrt{c^2 - v^2}$.

We can observe particles with short lifetimes in cosmic rays from distant places (several light years) [11]. This is because particles move at high velocities (close to the speed of light) and time slows down in the coordinates co-moving with the particles.

The velocity component (V_x, V_y, V_z) in the moving coordinate system is given by:

$$V'_x = \frac{dx'}{dt'} = \frac{d[\gamma(x - vt)]}{dt} \frac{dt}{d\left[\gamma\left(t - \frac{vx}{c^2}\right)\right]} \quad \left(\gamma = \frac{1}{\sqrt{1-\left(\frac{v}{c}\right)^2}}\right)$$

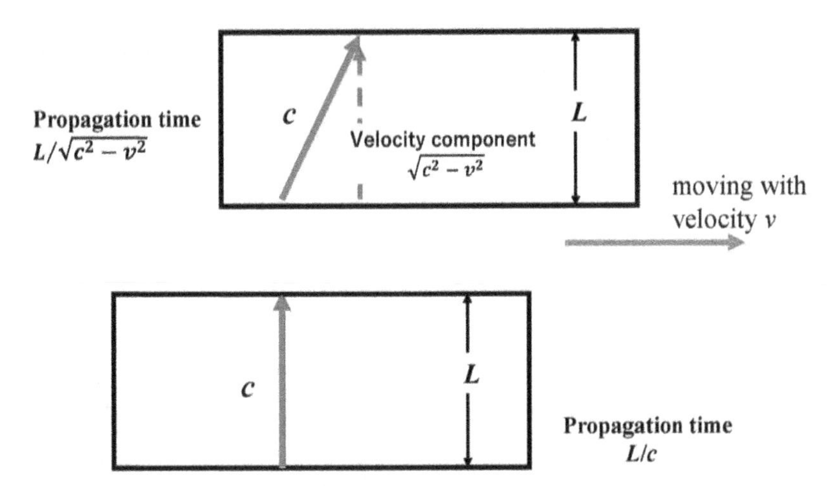

Figure 1.6. Time goes slower in a moving frame, considering that the light propagation time becomes longer.

taking $\frac{\mathrm{d}x}{\mathrm{d}t} = V_x$

$$= \frac{\mathrm{d}[\gamma(x - vt)]}{\mathrm{d}\left[\gamma\left(t - \frac{vV_x}{c^2}t\right)\right]} = \frac{V_x - v}{1 - \frac{vV_x}{c^2}}$$

$$V'_y = \frac{\mathrm{d}y'}{\mathrm{d}t'} = \frac{\mathrm{d}y}{\mathrm{d}\left[\gamma\left(t - \frac{vV_x}{c^2}t\right)\right]} = \frac{1}{\gamma}\frac{V_y}{1 - \frac{vV_x}{c^2}}$$

$$V'_z = \frac{1}{\gamma}\frac{V_z}{1 - \frac{vV_x}{c^2}}. \tag{1.3.9}$$

Here, we consider the case $V_x = c\cos\theta$ and $V_y = c\sin\theta$. Then

$V'_x = \frac{c\cos\theta - v}{1 - \frac{v}{c}\cos\theta}$, $V'_y = \sqrt{1 - \left(\frac{v}{c}\right)^2}\frac{c\sin\theta}{1 - \frac{v}{c}\cos\theta}$

$$V'^2_x + V'^2_y = c^2 \tag{1.3.10}$$

and the speed of light is constant for any moving coordinate system.

The conservation of momentum and energy E ($\overrightarrow{p_1} + \overrightarrow{p_2} = \overrightarrow{p_3} + \overrightarrow{p_4}$, $E_1 + E_2 = E_3 + E_4$) is fundamental to mechanics, and we assume that it holds in any coordinates. For classical mechanics, momentum $\vec{p}$ is defined as $\vec{p} = m\vec{V}$ (m: mass, $\vec{V}$: velocity). The conservation of momentum with this definition is assumed to hold in a coordinate. In another coordinate, the momentum should be given by $\overrightarrow{p'} = m\overrightarrow{V'}$ ($\overrightarrow{V'}$ is obtained from equation (1.3.9)), with which the conservation of momentum does not hold. We need another definition of momentum that converges to the momentum of classical mechanics for $|\vec{V}| \ll c$. If the four-dimensional momentum vectors $\overrightarrow{p^4} = \left(p_x, p_y, p_z, E/c\right)$ are transformed into a moving coordinate system using the Lorenz transformation:

$$\begin{pmatrix} p'x \\ p'y \\ p'z \\ E'/c \end{pmatrix} = \begin{pmatrix} \frac{1}{\sqrt{1-(v/c)^2}} & 0 & 0 & \frac{-(v/c)}{\sqrt{1-(v/c)^2}} \\ 0 & 1 & 0 & 0 \\ 0 & 0 & 1 & 0 \\ \frac{-(v/c)}{\sqrt{1-(v/c)^2}} & 0 & 0 & \frac{1}{\sqrt{1-(v/c)^2}} \end{pmatrix} \begin{pmatrix} p_x \\ p_y \\ p_z \\ E/c \end{pmatrix}, \tag{1.3.11}$$

the $\overrightarrow{p_1^4} + \overrightarrow{p_2^4} = \overrightarrow{p_3^4} + \overrightarrow{p_4^4}$ and $\overrightarrow{p_1^{4\prime}} + \overrightarrow{p_2^{4\prime}} = \overrightarrow{p_3^{4\prime}} + \overrightarrow{p_4^{4\prime}}$ are equivalent. Therefore, the conservation of the total momentum and energy are guaranteed in any coordinate system. We consider the transformation from $\overrightarrow{p_4} = (0,0,0, E/c)$,

$$p_x' = \frac{-\frac{v}{c^2}E}{\sqrt{1-\left(\frac{v}{c}\right)^2}} = \frac{\frac{V_x'}{c^2}E}{\sqrt{1-\left(\frac{V_x'}{c}\right)^2}} \tag{1.3.12}$$

$$E' = \frac{E}{\sqrt{1-\left(\frac{V_x'}{c}\right)^2}}. \tag{1.3.13}$$

Assuming the rest energy of $E = mc^2$,

$$p_x' = \frac{mV_x'}{\sqrt{1-\left(\frac{V_x'}{c}\right)^2}}, \tag{1.3.14}$$

$$E' = \frac{mc^2}{\sqrt{1-\left(\frac{V_x'}{c}\right)^2}}. \tag{1.3.15}$$

For $V_x' \ll c$, equations (1.3.14) and (1.3.15) are approximated to:

$$\begin{aligned} p_x' &= mV_x' \\ E' &= mc^2 + \frac{m}{2}V_x'^2. \end{aligned} \tag{1.3.16}$$

Equation (1.3.16) shows the momentum for classical mechanics and the energy is the sum of the rest energy and kinetic energy in the case of classical mechanics. Here we confirm that the momentum and energy defined by equations (1.3.14) and (1.3.15) is valid also after the Lorenz transform shown equation (1.3.11). The Lorenz transform from $\left(p_x, 0,0, E/c\right)$ is given by

$$p_x' = \frac{p_x - vE/c^2}{\sqrt{1-(v/c)^2}}$$

$$E'/c^2 = \frac{E/c^2 - vp_x/c^2}{\sqrt{1 - (v/c)^2}}. \tag{1.3.17}$$

Using equations (1.3.14) and (1.3.15)

$$p_x = \frac{mV_x}{\sqrt{1 - \frac{V_x^2}{c^2}}},\quad E' = \frac{mc^2}{\sqrt{1 - \frac{V_x^2}{c^2}}}$$

$$p_x' = \frac{mV_x'}{\sqrt{1 - \frac{V_x'^2}{c^2}}} = \frac{m(V_x - v)}{\left(1 - \frac{V_x v}{c^2}\right)\sqrt{1 - \left[\frac{V_x - v}{c^2 - V_x v}\right]^2}} = \frac{m(V_x - v)}{\sqrt{1 - (V_x/c)^2}\sqrt{1 - (v/c)^2}} = \frac{p_x - vE/c^2}{\sqrt{1 - (v/c)^2}}$$

$$E' = \frac{mc^2}{\sqrt{1 - \frac{V_x'^2}{c^2}}} = \frac{mc^2}{\sqrt{1 - \left[\frac{V_x - v}{c^2 - V_x v}\right]^2}} = \frac{E/c^2 - vp_x/c^2}{\sqrt{1 - (v/c)^2}}$$

$$\left(\text{using}\quad V_x' = \frac{V_x - v}{1 - \frac{V_x v}{c^2}}\right), \tag{1.3.18}$$

and equation (1.3.17) was confirmed. Giving an accelerating force, the momentum and energy increase but the speed cannot exceed that of light.

Comparing equations (1.3.14) and (1.3.15),

$$E^2 = (mc^2)^2 + c^2 |\vec{p}|^2, \tag{1.3.19}$$

is derived. Equation (1.3.19) is also applicable to bodies with $m = 0$, for example a photon. Light gives a radiation pressure when it is reflected by a mirror, showing that light has a momentum of $\frac{E_{\text{light}}}{c}$ (E_{light}: light energy) in the propagation direction, which is consistent with equation (1.3.19). For the theory of special relativity, the relationship between momentum and energy is unified by equation (1.3.19) for all matters, with and without mass. All particles with $m = 0$ must always move at the speed of light and there is no temporal procedure (see equation (1.3.8))

When there is an electromagnetic field, the momentum and energy are transformed to $\vec{p} \rightarrow \vec{p} + q_e\vec{A}$ and $E \rightarrow E - q_e\Phi_{el}$ ($\vec{A}$: magnetic vector potential with which magnetic field is given by $\vec{B} = \nabla \times \vec{A}$, Φ_{el}: electric voltage with which the electric field is given by $\vec{E} = -\nabla\Phi_{el} - \partial\vec{A}/\partial t$, q_e: electric charge). Therefore, the following transformation is valid:

$$\begin{pmatrix} A_x' \\ A_y' \\ A_z' \\ \Phi'_{el}/c \end{pmatrix} = \begin{pmatrix} \frac{1}{\sqrt{1 - (v/c)^2}} & 0 & 0 & \frac{-(v/c)}{\sqrt{1 - (v/c)^2}} \\ 0 & 1 & 0 & 0 \\ 0 & 0 & 1 & 0 \\ \frac{-(v/c)}{\sqrt{1 - (v/c)^2}} & 0 & 0 & \frac{1}{\sqrt{1 - (v/c)^2}} \end{pmatrix} \begin{pmatrix} A_x \\ A_y \\ A_z \\ \Phi_{el}/c \end{pmatrix}. \tag{1.3.20}$$

Maxwell's equations are also valid for another moving coordinate system. However, the electric and magnetic fields are different from those of the other coordinate systems. When $\vec{A} = 0$ and $\Phi_{el} \neq 0$, there is no magnetic field in this coordinate. However, in the moving system $A'_x \neq 0$ and the magnetic field exists because the charged matter moves in another coordinate system.

The dependence of the electromagnetic field on the reference frame is considered using the following model. We consider an electric current as viewed by observer A as having a positive electric charge of density $+\rho$ moving with velocity $+V$ and a negative electric charge with density $-\rho$ moving with velocity $-V$. Although the electric field is zero, there is a magnetic field induced by the current of $2\rho V$. For observer B co-moving with the positive charge, the velocity of the negative charge is not $2V$, but $2V/(1 + V^2/c^2)$. Hence, the magnetic field from the perspective of observer B is smaller than that for observer A. Moreover, the unit of length for the negative charge is shorter, the density of the negative charge is higher than that of the positive charge, and the total electric charge density is $\rho\left[1 - \frac{1}{\sqrt{1-(2V/c)^2}}\right]$. Therefore, the electric field is non-zero.

1.4 Theory of general relativity

We consider the case that A and B are moving with a high relative velocity. With the theory of special relativity, the relativistic effect that A observes with B is same that B observes with A. For example, A observes that time goes slower with B, while also B observes that time goes slower with A. When they meet again, which is younger?

This question is meaningless because they cannot meet again without any acceleration, while the theory of special relativity is valid only in the inertial frame of reference. When only A or B accelerate, only one person gets the inertial force. Then the relativistic effect that A observes from B and B observes from A should not be equal. Development of the theory of general relativity was required to discuss the mechanics of acceleration.

The fundamentals of the theory of general relativity is the equivalence principle between the inertial force and the gravitational force [10]. Newtonian mechanics indicated that both inertial force and the gravitational force are proportional to the mass, but 'inertial mass' and 'gravitational mass' were treated as different (only empirically looked like the same one). With the theory of general relativity, both forces are the same. When we are in an accelerated system, we see that the optical path is bent (figure 1.7). Einstein expected that the bending of the optical path was caused also by gravity. Then the position of stars should be observed at a shifted position when the optical path is bent by the passing of a nearby massive star. This expectation was confirmed in 1919 by Eddington when observing a distant star during a solar eclipse [12]. There is also a phenomenon called 'gravitational lens', with which the light from distant stars is focused by massive materials on the path to the Earth, as shown in figure 1.8. The gravitational effect makes it possible to observe the light from a star in a distant place. We can measure the mass distribution in the Schrödinger using the gravitational lens effect.

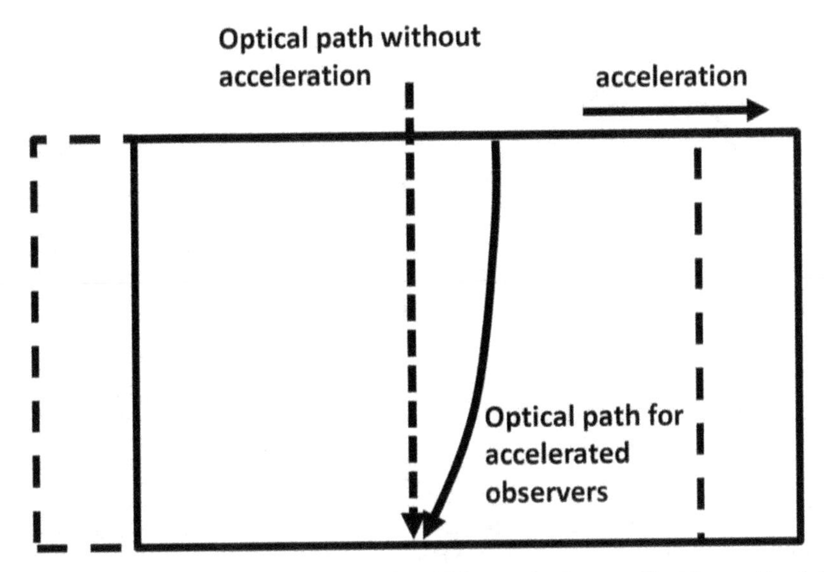

Figure 1.7. Bending of an optical path in an accelerated frame of reference. Einstein considered the same phenomenon also with gravity. Reproduced from [31]. Copyright IOP Publishing Ltd. All rights reserved.

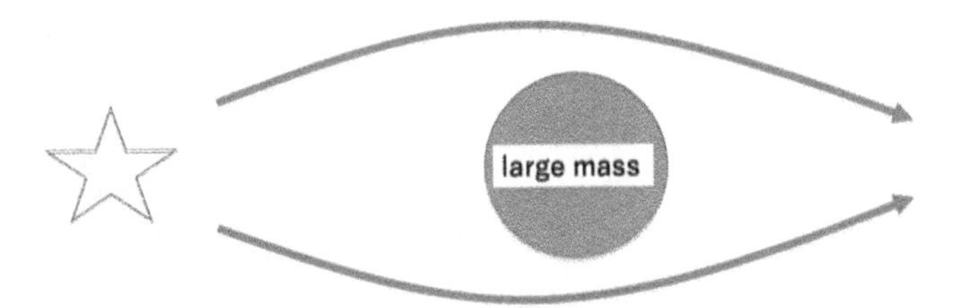

Figure 1.8. The gravitational lens effect. The light from a star can be focused by a massive object on the way to the Earth.

The bend in the optical path indicates the existence of a black hole, which is the region where gravity is so strong that nothing, including light, can escape its event horizon as shown in figure 1.9. No light can propagate from a black hole to other space. Black holes are observed as dark spots.

The length of the bent optical path is longer than the straight one. As the speed of light is an absolute constant, the propagation time is longer. Here we consider the propagation time of the light in the horizontal direction. The horizontal component of speed of light at time t ($t = 0$ is the irradiation time) is $\sqrt{c^2 - (gt)^2}$, where g is the gravitational acceleration. Considering the gravitational potential $\Phi_G = -(gt)^2/2$, the time goes slower

$$\Delta T'(\Phi_G) = \Delta T\sqrt{1 + \frac{2\Phi_G}{c^2}}\ \Phi_G < 0, \tag{1.4.1}$$

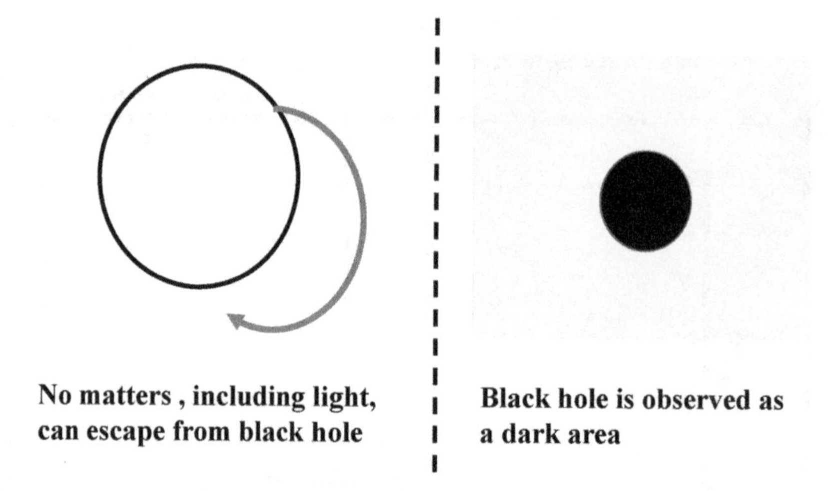

Figure 1.9. An imagination of a black hole. No matter, including light, can escape from the area where the gravitational potential energy is high enough. The black hole is observed as a dark area.

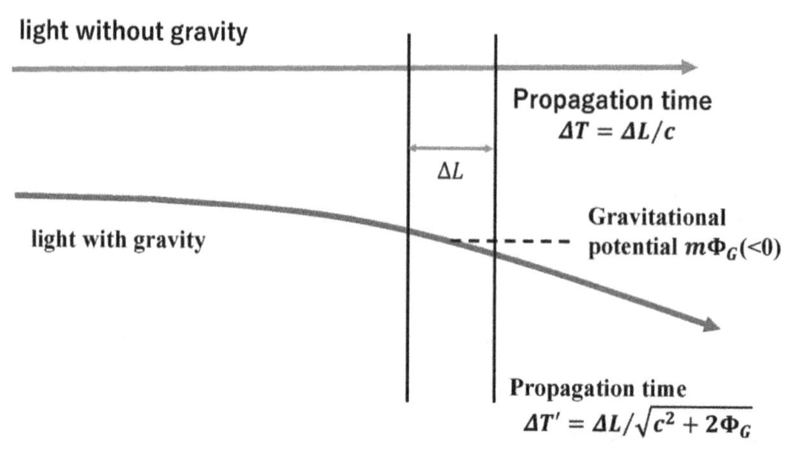

Figure 1.10. The concept of the gravitational red shift. In the gravitational potential field, the light path is bent and the propagation time becomes longer.

as shown in figure 1.10 (called gravitational red shift). We can understand this formula using the theory of special relativity with $mv^2 = 2\Phi_G$ from the change in kinetic energy induced by the acceleration due to gravity.

With Newtonian mechanics, gravity is an attractive force between matter having mass. However, gravity also bends optical paths although the mass of photon is zero. However, there is another problem, in that the phase of light cannot be uniform in the cross section with the constant speed while considering a straight coordinate. Einstein gave an idea that gravity distorts the space (bending the coordinate axis) and matter moves uniformly along the bent axis. When there is a temporal change of the gravitational potential, the distortion of space (change of the figure of coordinate axis) propagates as a wave with the speed of light, which is called a 'gravitational wave'. We can imagine a model with the object moving on a flexing mat (figure 1.11). Propagation with the speed of light can be imagined that the change of the optical path by the change of the gravitational potential is recognized after the propagation by light.

The gravitational wave propagating the z-direction gives the change of length of matter in the x- and y- directions by

$$
\begin{aligned}
L_x(t, z) &= L_{x0}(z)\left\{1 + \delta_G \sin\left[2\pi\nu_R\left(t - \frac{z}{c}\right)\right]\right\} \\
L_y(t, z) &= L_{y0}(z)\left\{1 - \delta_G \sin\left[2\pi\nu_R\left(t - \frac{z}{c}\right)\right]\right\},
\end{aligned}
\tag{1.4.2}
$$

where ν_R is the frequency of the revolution motion. The energy of revolution motion of binary stars decreases by the expansion of the gravitational wave and the distance between the stars get closer, which is observed as the increase of ν_R (section 2.10). The direct observation of the gravitational wave has been very difficult, because δ_G is below 10^{-21} also by the merging of black holes.

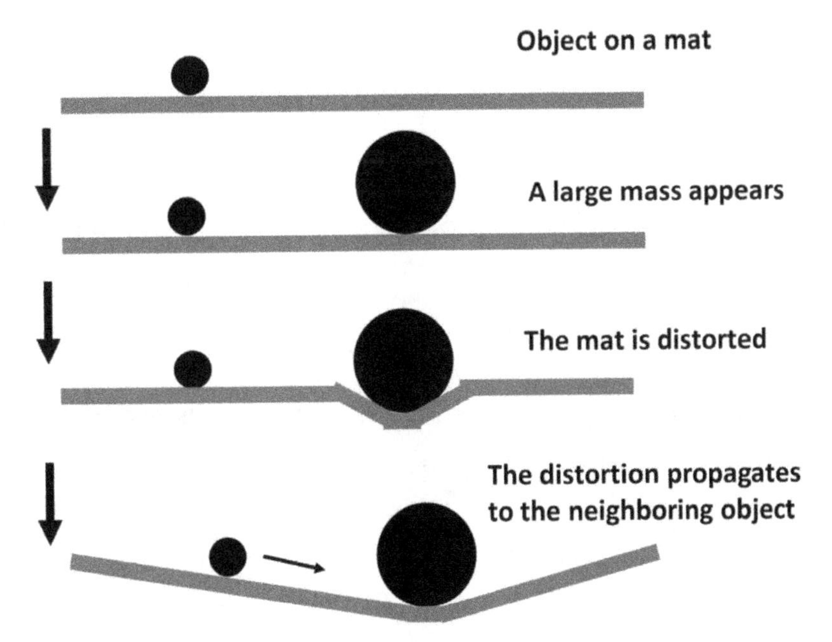

Figure 1.11. Space distortion induced by gravity can be imagined with the distortion of the surface of a flexing mat when a large object is placed on it. Reproduced from [31]. Copyright IOP Publishing Ltd. All rights reserved.

1.5 Quantum mechanics

1.5.1 Wave characteristics of all particles

Newtonian mechanics was valid while atoms were recognized as the smallest unit of matter. The situation was changed drastically by the discovery of electrons [13] and atoms were realized to have structures including electrons. Rutherford scattering [14] indicated that atoms are formed by a nucleus with positive electronic charge and electrons with negative charge revolving around the nucleus. Previous electromagnetism and Newtonian mechanics indicated that electrons under the electric field should lose their kinetic energy by irradiating an electromagnetic wave and finally merge to the nucleus. This discrepancy indicated that the motion of electrons cannot be described by previous physics.

There was another mystery that only the wavelength λ_{emit} satisfying

$$\frac{1}{\lambda_{emit}} = Ry\left[\frac{1}{n_1^2} - \frac{1}{n_2^2}\right] \quad n1,\ 2\text{: integer } Ry\text{: Rydberg constant} \tag{1.5.1}$$

were observed from the emission from hydrogen atoms [15, 16]. In 1913, Bohr established the 'old quantum mechanics' from the assumption that particles bounded in a limited area of q must satisfy [17]

$$\oint p_q dq = nh \quad n\text{: integer.} \tag{1.5.2}$$

From the balance between the centrifugal force and the Coulomb force,

$$\frac{\mu_e v^2}{r} = \frac{e^2}{4\pi\varepsilon_0 r^2}$$

e: unit electric charge

$$\mu_e \left(= \frac{m_e m_p}{m_e + m_p}\right)\text{: reduced mass between electron (mass: } m_e\text{) and proton (mass of } m_p\text{)} \tag{1.5.3}$$

From equations (1.5.2) and (1.5.3), possible radius of electron orbit is given by

$$r = a_B n^2$$
$$a_B = \frac{\varepsilon_0 h^2}{\pi \mu_e e^2}, \tag{1.5.4}$$

where a_B is called Bohr radius. The possible electron energy is given by

$$E_{en} = -\frac{e^2}{8\pi\varepsilon_0 a_B}\frac{1}{n^2} \tag{1.5.5}$$

When electron energy changes, the light with the frequency of

$$\nu_{n,\,n'} = \frac{E_{en} - E_{en'}}{h} \tag{1.5.6}$$

is absorbed or emitted, so that the sum of the electron energy and the photon energy is conserved. This estimation is consistent with the experimental results shown in equation (1.5.1), taking Rydberg constant,

$$Ry = \frac{\mu_e e^4}{8\varepsilon_0^2 h^3 c}. \tag{1.5.7}$$

Because of the discrete energy level, electron can stay in the orbits.

The idea of a matter wave was given by de Broglie [18]. Not only light, but all particles have the dual characteristic (particle and wave). The energy of each particle is $E = h\nu$ and the momentum is $\vec{p} = h\vec{k}$ $(|\vec{k}| = 1/\lambda)$, where ν is the frequency, $\vec{k}$ is the wavenumber vector and λ is the wavelength (this relation was confirmed with light). From the relation $|\vec{p}| = h/\lambda$, equation (1.5.2) is interpreted as the description of the resonant wavelength regulated by the bounding state. The particles in a limited space area can take only discrete energies because the waves can take only resonant wavelength, which determines the kinetic energy.

1.5.2 Fundamental of quantum mechanics

With quantum mechanics, properties of all particles are given using wavefunction Φ. The wavefunction of sinusoidal oscillation is often given using a complex function $\exp(ix) = e^{ix} = \cos x + i \sin x$, because the absolute value does not change with the oscillation $(|\exp(ix)| = 1)$. Note that the optical energy density (density of the photon number) is proportional to the square of the amplitude of the electromagnetic wave. In quantum mechanics, the properties of all particles are given using wavefunction Φ, and the existence probability is given by $|\Phi|^2$, in analogy with the relation between the photon density and the oscillation amplitude of the electromagnetic field. The physical value X is given by

$$\int \Phi^* \overline{X} \Phi d\vec{r} \tag{1.5.8}$$

as the average for the given state. Here, $\overline{X}$ is the operator for X. Wavefunctions having deterministic values of X (called eigenvalue X_e) are called eigenfunctions Φ_e, which satisfy

$$\check{X}\Phi_e = X_e\Phi_e. \tag{1.5.9}$$

All functions can be expressed as a linear combination of eigenfunctions as follows:

$$\Phi = \sum a_i \Phi_{ei}$$

$$\int \Phi_{ei}^* \Phi_{ei} d\vec{r} = 1$$

$$\int \Phi_{ei}^* \Phi_{ej} d\vec{r} = 0 \ (i \neq j)$$

$$\sum | a_i |^2 = 1. \tag{1.5.10}$$

The measurement result of X is one of the eigenvalues X_{ei} with a probability of $| a_i |^2$. Owing to the measurement, the wavefunction changes to the eigenfunction corresponding to the measured eigenvalue. The change in the quantum state by the measurement is called 'quantum destruction'. Values of a_i are obtained by

$$a_i = \int \Phi_{ei}^* \Phi \mathrm{d}\vec{r}. \tag{1.5.11}$$

Here we consider the relation between the eigenfunctions of $X(\Phi_{ei})$ and $Y(\Phi'_{ei})$,

$$\begin{aligned} \Phi'_{ei} &= \sum a_{ji}\Phi_{ej} \; a_{ji} = \int \Phi_{ej}^* \Phi'_{ei} \mathrm{d}\vec{r} \\ \Phi_{ei} &= \sum b_{ji}\Phi'_{ej} \; b_{ji} = \int \Phi_{ej}^{'*} \Phi_{ei} \mathrm{d}\vec{r} \\ b_{ij} &= a_{ji}^* \end{aligned} \tag{1.5.12}$$

The product of matrixes a_{ij} and b_{ij} should be the unit matrix. Therefore, these matrixes are unitary matrixes.

Here, we discuss the possibility to get the deterministic two physical values simultaneously. The measurement uncertainties of X and Y are given by $(\check{X} - X_e)^2$ and $(\check{Y} - Y_e)^2$. With the operators, we assume $\check{X}\check{Y} - \check{Y}\check{X} = \delta$. Then

$$[(\check{X} - X_e) - i(\check{Y} - Y_e)][(\check{X} - X_e) + i(\check{Y} - Y_e)] > 0$$

$$(\check{X} - X_e)^2 + (\check{Y} - Y_e)^2 + i(\check{X}\check{Y} - \check{Y}\check{X}) > 0$$

$$\min[(\Delta X)^2 + (\Delta Y)^2] = 2\Delta X \Delta Y > | \delta | \;\; \Delta X = | X - X_e | \;\; \Delta Y = | Y - Y_e | \tag{1.5.13}$$

Then X and Y cannot be deterministic simultaneously when $\delta \neq 0$ and the relation of uncertainties of both values is given by $\Delta X \Delta Y > | \delta |/2$.

With deterministic values of energy E and momentum $p_{x,y,z}$, the wavefunction is given by

$$\Phi = \exp\left[i\frac{2\pi}{h}\left(Et + p_x x + p_y y + p_z z\right)\right] \tag{1.5.14}$$

and

$$H\Phi = \frac{h}{2\pi i}\frac{\partial \Phi}{\partial t} = E\Phi, \; \frac{h}{2\pi i}\frac{\partial \Phi}{\partial q} = p_q \Phi \;\; q = x, y, z, \tag{1.5.15}$$

is derived (h: Planck constant). Here, H is the operator called Hamiltonian. Note

$$Ht - tH = p_q q - q p_q = \frac{h}{2\pi i}. \tag{1.5.16}$$

Equations (1.5.13)–(1.5.16) show

$$\Delta E \Delta t > \frac{h}{4\pi}, \ \Delta p_q \Delta q > \frac{h}{4\pi}. \tag{1.5.17}$$

The phase of the wavefunction given by Et/h and $p_q q/h$ has the uncertainty of the order of $\pm 1/2$ radian. With a deterministic E and p_q, $|\Phi|^2$ with the wave given by equation (1.5.14) is totally uniform in the q-direction and there is no temporal change. When p_q distribute uniformly with $-p_a < p_q < p_a$,

$$\Phi = \int_{-p_a}^{p_a} \exp\left(\frac{2\pi i p_q q}{h}\right) dp_q = \frac{h}{2\pi i q}\left[\exp\left(\frac{2\pi i p_a q}{h}\right) - \exp\left(-\frac{2\pi i p_a q}{h}\right)\right] = \frac{h}{\pi q}\sin\left(\frac{2\pi p_a q}{h}\right)$$

$$|\Phi|^2 = \left[\frac{h}{\pi}\frac{\sin\left(\frac{2\pi p_a q}{h}\right)}{q}\right]^2. \tag{1.5.18}$$

$|\Phi|^2$ is zero at $q = \pm h/2p_a$ and the particle is localized in the area $-h/2p_a < q < h/2p_a$. As the momentum distribution becomes larger, the particle can localize in a smaller area. The uncertainty principle indicates the position uncertainty (broadening of wave packet) is the order of $\Delta q = h/\left(4\pi \Delta p_q\right)$. Using the kinetic temperature T_K (temperature indicating the broadening of the kinetic energy distribution), $\Delta p_q = \sqrt{m k_B T_K}$ (m: mass, k_B: Boltzmann constant 1.38×10^{-23} J K^{-1}) and

$$\Delta q = \frac{h}{4\pi\sqrt{m k_B T_K}}. \tag{1.5.19}$$

The broadening of the wave packet is significant with lower kinetic temperature and smaller mass. With $T_K = 300$K, Δq is 2×10^{-12} m with ^{87}Rb atom, which is negligibly small in comparison with the radius of this atom. Therefore, the motion of the atom is described well with Newtonian mechanics. With $T_K = 1 \times 10^{-6}$K, it is broadened up to 4×10^{-8} m. With electron, it is 8.5×10^{-10} m with $T_K = 300$K and quantum mechanical treatment is required with electrons also with room temperature.

The uncertainty principle was established without distinguishing the limit of the measurement uncertainty $\Delta\epsilon_m$, scattering by the measurement $\Delta\eta_s$, and the quantum fluctuation $\Delta\sigma_Q$. Ozawa insisted that these uncertainty components should be distinguished, and the uncertainty principle should be corrected to [19]

$$\Delta\epsilon_m\left(\Delta\eta_s + \Delta\sigma_Q\right) + \Delta\eta_s \Delta\sigma_Q \geqslant \frac{h}{4\pi}. \tag{1.5.20}$$

Then the measurement uncertainty can be

$$\Delta\epsilon_m < \frac{h}{4\pi\left(\Delta\eta_s + \Delta\sigma_Q\right)}, \tag{1.5.21}$$

which was experimentally confirmed by the measurement of the nuclear spin (spin is introduced in section 1.6) [20].

The motion of matter is described as the temporal changes of the density distribution. As shown above, the localization of wave distribution and its temporal change is possible only when the wavefunctions of different momentum and energy are coupled. We consider the temporal change in the density distribution from the interference between the two components of energy and momentum as follows:

$$\Phi = \exp\left[\frac{2\pi i}{h}\left(E_1 t + p_{q1} q\right)\right] + \exp\left[\frac{2\pi i}{h}\left(E_2 t + p_{q2} q\right)\right]$$

$$\Phi^* = \exp\left[-\frac{2\pi i}{h}\left(E_1 t + p_{q1} q\right)\right] + \exp\left[-\frac{2\pi i}{h}\left(E_2 t + p_{q2} q\right)\right]$$

(using $\exp(ix) = \cos(x) + i\sin(x)$)

$$|\Phi|^2 = 2 + 2\cos\left[\frac{2\pi}{h}(E_1 - E_2)t + \frac{2\pi}{h}(p_{q1} - p_{q2})q\right]. \tag{1.5.22}$$

The motion velocity of the particle corresponds to the temporal change of the probability distribution given by $v_q = \frac{E_1 - E_2}{p_{q1} - p_{q2}} = \frac{dE}{dp_q}$ using $E = \frac{[p]^2}{2m}$

$$v_q = \frac{p_q}{m}, \tag{1.5.23}$$

which corresponds to the definition of momentum in Newtonian mechanics. The velocity given by equation (1.5.23) is called the 'group velocity' which is quite different from the velocity of the phase propagation E/p_q. Considering the rest energy mc^2 (derived by special theory of relativity), the velocity of the phase propagation is higher than the speed of light. However, it does not influence physical phenomena and the theory of relativity (prohibiting a velocity higher than speed of light) is not violated. Equation (1.5.23) is derived also by the relativistic treatment (equation (1.3.19)) with the approximation with $cp \ll mc^2$. For another observer moving with a velocity of v_q', the observed frequency shifts by the Doppler effect ($E/h \to (E + \Delta E)/h$). The observed wavelength $\left(=h/p_q\right)$ does not change, ignoring the relativistic effect. The group velocity is given by

$$\frac{E + \Delta E}{h} = \frac{E}{h}\left(1 - \frac{p_q v_q'}{E}\right) = \frac{E - p_v v_q'}{h}$$

$$\frac{\mathrm{d}[E + \Delta E]}{\mathrm{d}p_q} = \frac{\mathrm{d}E}{\mathrm{d}p_q} - v_q' = v_q - v_q'. \tag{1.5.24}$$

The Schrödinger equation for particles with a mass of m is derived from the relation between the energy and momentum given by Newtonian mechanics

$$E = \frac{p_x^2 + p_y^2 + p_z^2}{2m} + V(x, y, z) \quad V(x, y, z)\text{: potential energy,} \tag{1.5.25}$$

to be

$$H\Phi = \frac{h}{2\pi i}\frac{\partial \Phi}{\partial t} = \left[-\frac{h^2}{8\pi^2 m}\left(\frac{\partial^2}{\partial x^2} + \frac{\partial^2}{\partial y^2} + \frac{\partial^2}{\partial z^2}\right) + V(x, y, z)\right]\Phi. \tag{1.5.26}$$

When Φ is an energy eigenfunction, equation (1.5.26) is rewritten as

$$\begin{aligned} \Phi &= \exp\left[\frac{2\pi i}{h}Et\right]\Psi(x, y, z) \\ E\Psi &= \left[-\frac{h^2}{8\pi^2 m}\left(\frac{\partial^2}{\partial x^2} + \frac{\partial^2}{\partial y^2} + \frac{\partial^2}{\partial z^2}\right) + V(x, y, z)\right]\Psi. \end{aligned} \tag{1.5.27}$$

In free space, the one-dimensional Schrödinger equation is

$$H\Psi = -\frac{h^2}{8\pi^2 m}\frac{\partial^2}{\partial x^2}\Psi = E\Psi \tag{1.5.28}$$

and its solution is

$$\begin{aligned} \Psi &= c_+ \exp\left(\frac{2\pi i\sqrt{2mE}}{h}x\right) + c_- \exp\left(-\frac{2\pi i\sqrt{2mE}}{h}x\right) \\ &= C_p \sin\left[\frac{2\pi\sqrt{2mE}}{h}x + \delta_p\right]. \end{aligned} \tag{1.5.29}$$

Here we consider a potential well ($V(x) = 0$ at $0 < x < L$, and $V(x) = \infty$ at $x < 0$ and $x > L$). The solution for $0 < x < L$ is given by equation (1.5.29) with the requirement of $\varphi = 0$ at $x = 0, L$. Therefore,

$$\begin{aligned} \Psi &\propto \sin\left[\frac{2\pi\sqrt{2mE_{n_t}}}{h}x\right] = \sin\left[\frac{n_t \pi}{L}x\right] \quad n_t\text{: integer} \\ E_{n_t} &= \frac{1}{2m}\left(\frac{hn_t}{2L}\right)^2 \end{aligned} \tag{1.5.30}$$

is obtained. The non-zero lowest energy is interpreted from the uncertainty principle between momentum and position. Limiting the position region as $0 < x < L$, the uncertainty of the momentum larger than $h/4\pi L$ is required. As a matter is localized in a narrower space area, the gap between possible eigen energies becomes larger. The wavefunction is the standing wave, without a propagation direction. This wavefunction is the sum of the traveling waves in the positive and negative directions; in other words, coupling of momentums $p_x = \pm hn_t/2L$.

When the potential is given by a central force (depend only $r = \sqrt{x^2 + y^2 + z^2}$), it is convenient to use the polar coordinate (r,θ,φ), $x = r\sin\theta\cos\varphi$, $y = r\sin\theta\sin\varphi$, $z = r\cos\theta$. Then equation (1.5.26) is rewritten as

$$H = \left[-\frac{h^2}{8\pi^2 m}\left(\frac{\partial^2}{\partial r^2} + \frac{2}{r}\frac{\partial}{\partial r}\right) + \frac{1}{2mr^2}\widetilde{L^2} + V(r)\right],$$

$$\check{L}_x = \frac{h}{2\pi i}\left[-\sin\varphi\frac{\partial}{\partial\theta} - \frac{\cos\theta}{\sin\theta}\cos\varphi\frac{\partial}{\partial\varphi}\right],$$

$$\check{L}_y = \frac{h}{2\pi i}\left[\cos\varphi\frac{\partial}{\partial\theta} - \frac{\cos\theta}{\sin\theta}\sin\varphi\frac{\partial}{\partial\varphi}\right],$$

$$\widetilde{L_z} = \frac{h}{2\pi i}\frac{\partial}{\partial\varphi},$$

$$\check{L}^2 = \check{L}_x^{\,2} + \check{L}_y^{\,2} + \check{L}_z^{\,2} = -\left(\frac{h}{2\pi}\right)^2\left[\frac{\partial^2}{\partial\theta^2} + \frac{\cos\theta}{\sin\theta}\frac{\partial}{\partial\theta} + \frac{1}{(\sin\theta)^2}\frac{\partial^2}{\partial\varphi^2}\right], \qquad (1.5.31)$$

where $\check{L}_q$ is the operator of the angular momentum by the rotation around the q-axis ($q = x, y, z$). Here we assume that the wavefunction is given by the formula of

$$\Psi(r, \theta, \phi) = R(r)Y(\theta)\Theta(\varphi). \qquad (1.5.32)$$

The eigenfunction and eigenvalue of L_z are given by

$$\Theta(\varphi) = \frac{1}{\sqrt{2\pi}}\exp(iM_L\varphi)$$

$$L_z = \frac{h}{2\pi}M_L, \qquad (1.5.33)$$

where M_L is integer so that $\Theta(\varphi + 2\pi) = \Theta(\varphi)$ is satisfied. M_L s called 'magnetic quantum number'. The absolute value of the angular momentum is given by a rotational quantum number L, and M_L are the integers between $\pm L$. From the uncertainty principle between φ and L_z, the distribution of $|\Theta|^2$ is perfectly uniform with φ with a deterministic value of L_z. We cannot determine the direction of the xy-axis and $L_{x,y}$ cannot be determined except for when $L_x = L_y = L_z = 0$. Here, we consider operators

$$\check{L}_\pm = \check{L}_x \pm i\check{L}_y = \frac{h}{2\pi i}\exp(\pm i\varphi)\left[\pm i\frac{\partial}{\partial\theta} - \frac{\cos\theta}{\sin\theta}\frac{\partial}{\partial\varphi}\right]. \qquad (1.5.34)$$

Then

$$\check{L}_\pm\Theta_{M_L}(\varphi) \propto \exp[i(M_L \pm 1)\varphi] \propto \Theta_{M_L\pm 1}(\varphi) \qquad (1.5.35)$$

and

$$\check{L}^2 = \check{L}_{\mp}\check{L}_{\pm} + \check{L}_z^{\ 2} \pm \frac{h}{2\pi}\check{L}_z \tag{1.5.36}$$

are obtained. To prohibit the wavefunction of $M_L = \pm(L + 1)$, there is a requirement

$$\check{L}_{\pm}Y_L^{\pm L}(\theta)\Theta_{\pm L}(\varphi) = 0. \tag{1.5.37}$$

The eigenvalue of the absolute value of the angular momentum is independent of M_L. With $M_L = \pm L$,

$$\check{L}^2 Y_L^{\pm L}(\theta)\Theta_{\pm L}(\varphi) = \left(\frac{h}{2\pi}\right)^2 L(L + 1)Y_L^{\pm L}(\theta)\Theta_{\pm L}(\varphi) \tag{1.5.38}$$

is obtained from equations (1.5.36) and (1.5.37). The square of the absolute value of the angular momentum is $(h/2\pi)^2L(L + 1)$.

Using equation (1.5.37),

$$\check{L}_{\pm}Y_L^{\pm L}(\theta)\Theta_{\pm L}(\varphi) = \frac{h}{2\pi i}\exp(\pm i\varphi)\left[\pm i\frac{\partial}{\partial\theta} - \frac{\cos\theta}{\sin\theta}\frac{\partial}{\partial\varphi}\right]Y_L^{\pm L}(\theta)\Theta_{\pm L}(\varphi) = 0$$

$$\pm i\frac{\partial}{\partial\theta}Y_L^{\pm L}(\theta) = \pm iL\frac{\cos\theta}{\sin\theta}Y_L^{\pm L}(\theta)$$

$$\frac{1}{Y_L^{\pm L}(\theta)}\frac{\partial}{\partial\theta}Y_L^{\pm L}(\theta) = L\frac{\cos\theta}{\sin\theta}$$

$$\int\frac{1}{Y_L^{\pm L}(\theta)}\mathrm{d}Y_L^{\pm L}(\theta) = L\int\frac{\cos\theta}{\sin\theta}\mathrm{d}\theta = L\int\frac{1}{\sin\theta}\mathrm{d}\{\sin\theta\}$$

$$Y_L^{\pm L}(\theta) \propto (\sin\theta)^L$$

$$Y_L^{\pm(L-n)}(\theta) \propto (\check{L}_{\mp})^n Y_L^{\pm L}$$

$$Y_L^{\pm(L-1)}(\theta) \propto (\sin\theta)^{L-1}\cos\theta$$

$$Y_0^0(\theta) = \sqrt{\frac{1}{2}},\ Y_1^0(\theta) = \sqrt{\frac{3}{2}}\cos\theta,\ Y_1^{\pm 1}(\theta) = \frac{\sqrt{3}}{2}\sin\theta,$$

$$Y_2^0(\theta) = \frac{1}{2}\sqrt{\frac{5}{2}}[3(\cos\theta)^2 - 1],\ Y_2^{\pm 1}(\theta) = \frac{\sqrt{15}}{2}\sin\theta\cos\theta,\ Y_2^{\pm 2}(\theta) = \frac{\sqrt{15}}{4}(\sin\theta)^2 \tag{1.5.39}$$

$R(r)$ is obtained using

$$ER(r) = \left[-\frac{h^2}{8\pi^2 m}\left(\frac{\partial^2}{\partial r^2} + \frac{2}{r}\frac{\partial}{\partial r}\right) + \frac{h^2L(L + 1)}{8\pi^2 mr^2} + V(r)\right]R(r). \tag{1.5.40}$$

Here we consider the energy of electron in a hydrogen atom taking $m \rightarrow \mu_e = \frac{m_e m_p}{m_e + m_p}$
$m_{e,p}$: mass of electron and proton

$$V(r) \rightarrow -\frac{e^2}{4\pi\varepsilon_0 r}. \tag{1.5.41}$$

Using the principal quantum number n (integer $n \geqslant 1$) the energy eigenvalues are obtained to be (detailed derivation is given in appendix B)

$$E_n = -\frac{e^2}{8\pi\varepsilon_0 a_B}\frac{1}{n^2} \tag{1.5.42}$$

which agrees with equation (1.5.5). However, the interpretation is quite different from the 'old quantum mechanics'. The electron motion in the $L = 0$ state is not a revolution, but the vibration in the radial direction. The ($n = 1$, $L = 0$) state is interpreted as the broadening of the electron wavefunction given by the uncertainty principle between the momentum and position in the radial direction. Electron quantum states are given by (n, L, M_L), where $0 \leqslant L < n$ and $-L \leqslant M_L \leqslant L$. Energy eigenvalues derived from Schrödinger equation are determined only by n. However, the actual energy structure is more complicated, because of the electron spin and relativistic effects as shown in section 1.6.

When there is a perturbation energy term (e.g, electromagnetic field) described by the Hamiltonian H', the total Hamiltonian is given by

$$\begin{aligned} H\Psi &= E_{eigen}\Psi \\ H &= H_0 + H'. \end{aligned} \tag{1.5.43}$$

The eigenfunction is given by the linear combination of eigenfunctions of H_0, as shown by equation (1.5.10). For simplicity, we consider with the coupling of two states,

$$\begin{aligned} \Psi &= a\Psi_a + b\Psi_b \\ H_0\Psi_{a,\,b} &= E_{a,\,b}\Psi_{a,\,b}. \end{aligned} \tag{1.5.44}$$

Using (1.5.43)–(1.5.44),

$$\begin{aligned} &\Psi_a^*(H_0 + H')(a\Psi_a + b\Psi_b) = a(E_a + H'_{aa}) + bH'_{ab} = aE_{eigen} \\ &\Psi_b^*(H_0 + H')(a\Psi_a + b\Psi_b) = aH'_{ba} + b(E_b + H'_{bb}) = bE_{eigen} \\ &H'_{ji} = \int \Psi_j^* H' \Psi_i d\vec{r} (H'^*_{ij} = H'_{ji}). \end{aligned} \tag{1.5.45}$$

To obtain non-zero coefficients a and b,

$$\begin{vmatrix} E_a + H'_{aa} - E_{eigen} & H'_{ab} \\ H'_{ba} & E_b + H'_{bb} - E_{eigen} \end{vmatrix} = 0 \tag{1.5.46}$$

is required. The energy eigenvalue is given by

$$E_{eig-u} = \frac{(E_a + H'_{aa} + E_b + H'_{bb}) + \sqrt{(E_a + H'_{aa} - E_b - H'_{bb})^2 + 4\,|\,H'_{ab}|^2}}{2}$$

$$E_{eig-d} = \frac{(E_a + H'_{aa} + E_b + H'_{bb}) - \sqrt{(E_a + H'_{aa} - E_b - H'_{bb})^2 + 4\,|\,H'_{ab}|^2}}{2}. \quad (1.5.47)$$

Assuming $E_b + H'_{bb} > E_a + H'_{aa}$ and $E_b + H'_{bb} - E_a - H'_{aa} \gg |\,H'_{ab}\,|$, equation (1.5.47) is approximated as

$$E_{eig-u} = E_b + H'_{bb} + \frac{|\,H'_{ab}|^2}{E_b + H'_{bb} - E_a - H'_{aa}},$$

$$E_{eig-d} = E_a + H'_{aa} - \frac{|\,H'_{ab}|^2}{E_b + H'_{bb} - E_a - H'_{aa}}. \quad (1.5.48)$$

When $H'_{ab} \neq 0$,

$$E_{eig-u} - E_{eig-d} = \sqrt{(E_a + H'_{aa} - E_b - H'_{bb})^2 + 4\,|\,H'_{ab}|^2} > 0. \quad (1.5.49)$$

Here, we discuss the 'anti-crossing', which is caused when $E_B - E_A > 0$ and $H'_{aa} - H'_{bb} > 0$. The eigenfunction corresponding to the E_{eig-u} and E_{eig-d} are denoted by Ψ_{eig-u} and Ψ_{eig-d}, respectively. The approximation of energy eigenvalues and eigenfunctions are shown below.

$$E_b - E_a \gg H'_{aa} - H'_{bb}$$

$$E_{eig-u} - E_{eig-d} \approx (E_b - E_a) - (H'_{aa} - H'_{bb})$$

$$\Psi_{eig-u} \approx \Psi_b$$

$$\Psi_{eig-d} \approx \Psi_a$$

$$E_b - E_a = H'_{aa} - H'_{bb}$$

$$E_{eig-u} - E_{eig-d} \approx 2\,|\,H'_{ab}|$$

$$\Psi_{eig-u} \approx \frac{\Psi_a \pm \Psi_b}{\sqrt{2}}$$

$$\Psi_{eig-d} \approx \frac{\Psi_a \mp \Psi_b}{\sqrt{2}}$$

$$E_b - E_a \ll H'_{aa} - H'_{bb}$$

$$E_{eig-u} - E_{eig-d} \approx (H'_{aa} - H'_{bb}) - (E_b - E_a)$$

$$\Psi_{eig-u} \approx \Psi_a$$

$$\Psi_{eig-d} \approx \Psi_b. \tag{1.5.50}$$

The transition between the a and b state can be performed by changing the interaction energy H' adiabatically (taking time much longer than $h/(2\pi \mid H'_{ab} \mid)$) so that the energy uncertainty is smaller than the energy gap between both states.

Here we consider the perturbation by the electric voltage field. The Hamiltonian of the interaction between an electric field $\vec{E}$ and electric dipole moment $\vec{d}$ is given by

$$\begin{aligned} H'_{ji} &= -\vec{d}_{ji} \cdot \vec{E} \\ \vec{d}_{ji} &= \int \rho_j^* \vec{r} \rho_i d\vec{r}, \end{aligned} \tag{1.5.51}$$

where ρ_i is the electric charge density distribution in each quantum state. The energy shift induced by the electric field is called 'Stark shift'. For atoms and linear molecules, $H'_{ii} = 0$ for the quantum state with a deterministic angular momentum, because the direction of the dipole moment is random from the uncertainty principle between angular momentum and direction. With a small electric field, the Stark shift is proportional to $\mid \vec{E} \mid^2$, which is called the 'quadratic Stark shift'. For non-linear polar molecules $H'_{ii} \neq 0$, and there is a linear Stark shift.

When there is magnetic field, the momentum is transformed to $\vec{p} \rightarrow \vec{p} + q_e\vec{A}$ ($\vec{A}$: magnetic vector potential satisfying $\vec{B} = \nabla \times \vec{A}$, q_e: electric charge) and equation (1.5.24) is rewritten as

$$H = \frac{1}{2m}\left(\vec{p} + q_e\vec{A}\right) \cdot \left(\vec{p} + q_e\vec{A}\right) + V \tag{1.5.52}$$

For $\vec{A} = \frac{1}{2}(-yB, xB, 0)$ and $\vec{B} = (0,0, B)$,

$$\frac{q_e\vec{p} \cdot \vec{A}}{m} = \frac{q_e B(xp_y - yp_x)}{2m} = \frac{q_e}{2m} L_z B = \frac{q_e h}{4\pi m} M_L B, \tag{1.5.53}$$

which is a term called the 'Zeeman energy shift'. $q_e L_z/2m$ is magnetic dipole moment. To describe the Zeeman energy shift in the electron energy state, a parameter called 'Bohr magneton $\mu_B = eh/4\pi m_e$' is often used. When the electron orbital angular momentum is the only angular momentum quantum number, the Zeeman energy shift is linear to the magnetic field strength, as shown in equation (1.5.53). There are different kinds of angular momentums like electron spin (section 1.6), nuclear spin, and molecular rotation. When there are coupling between different angular momentum states, the Zeeman shift should be obtained using the matrix of Hamiltonian-like equation (1.5.46).

1.5.3 The interference between undistinguishable phenomena

The most important characteristic of quantum mechanics is the interference between the undistinguishable phenomena. The simplest example is the interference of particles passing through a double slit (figure 1.12) [21]. We cannot distinguish

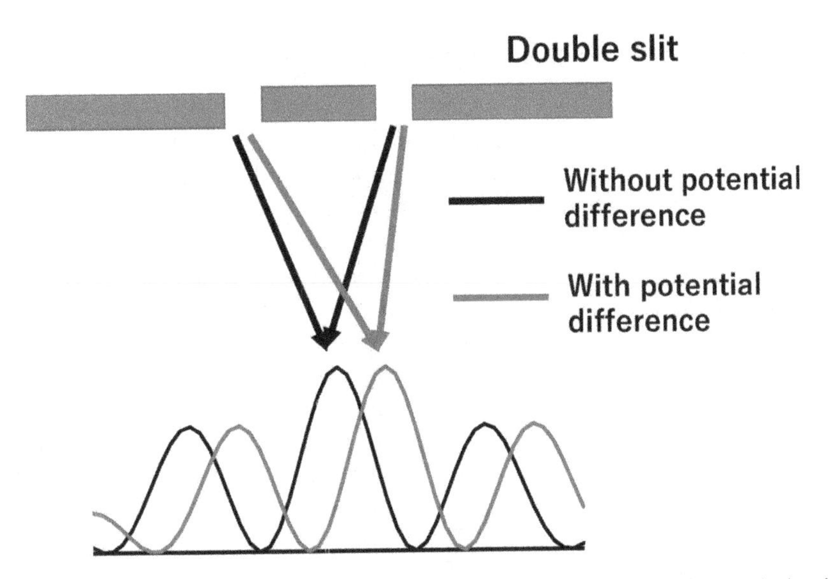

Figure 1.12. Interference of particles passing through a double slit. This interference is caused when it is not distinguishable between passing through both slits. When there is a potential difference between two paths, the interference signal is shifted. Reproduced from [32].

through which slit the particle passed. When there is a potential difference between two paths, the frequency difference between two paths is induced. This effect is observed as the shift of the interference signal.

There is an interference when the same two particles in the same quantum state exist. With this case, we cannot distinguish the relative position $\vec{r}$ or $-\vec{r}$. There are two kinds of particles, Boson and Fermion. The wavefunction with same two particles is

$$\begin{aligned} \Phi(\vec{r}) &\rightarrow \frac{\Phi(\vec{r}) + \Phi(-\vec{r})}{\sqrt{2}} \text{ for Boson particles} \\ \Phi(\vec{r}) &\rightarrow \frac{\Phi(\vec{r}) - \Phi(-\vec{r})}{\sqrt{2}} \text{ for Fermion particles.} \end{aligned} \tag{1.5.54}$$

For Fermions, only one particle with a single quantum state can exist in the same place because the wavefunction is zero with $\vec{r} = 0$. For example, the electron is Fermion and only one electron can exist in a (n, L, M_L, M_S) state, where M_S is the quantum number of the electron spin state which is introduced in section 1.6. Therefore, only two electrons can be in the ($n = 1$, $L = 0$) state. For a Boson, the square of the amplitude of wavefunction is double at $\vec{r} = 0$, which indicates that Boson particles tend to unify the quantum state. The discussion shown above is valid not only with exact $\vec{r} = 0$ but in the broadening of the wave packet (given by the uncertainty principle between position and momentum). Therefore, this effect is significant with low kinetic temperature. This characteristic gives significant effects shown in section 2.9.

1.6 Spin and relativistic quantum mechanics

1.6.1 Electron spin

The quantum energy state of an electron in an atom is given by the principal quantum number n, rotational quantum number L, and magnetic quantum number M_L, which are given by the motion of electrons under a Coulomb potential owing to the nucleus. An electron is a Fermion and only one electron can be in a quantum state as shown in section 1.5.3. However, two electrons can be in an (n, L, M_L) state, which means there are two states with electrons. In the Stern–Gerlach experiment, an Ag atomic beam passing through an area with inhomogeneous magnetic field was deflected from the straight path in two opposite directions. The deflection angle was the only quantized parameter [22]. The electron orbital angular momentum of the Ag atom was zero. This result shows that the electron has two states like a permanent magnet: the S-pole or N-pole in the direction of the magnetic field. In analogy with the M_L states for each L state (number of states $2L + 1$ with $-L \leqslant M_L \leqslant L$), these two states of electron were described as the virtual angular momentum states ($S = 1/2$, $M_S = \pm 1/2$) called spin. The spin state is a property of the electron itself and the eigenfunction is not derived from the density distribution in a potential field. Pauli proposed describing two spin states using a two-dimensional vector [23]. The eigenfunction of each spin state is as follows:

$$M_S = \frac{1}{2} \rightarrow \xi_+ = \begin{pmatrix} 1 \\ 0 \end{pmatrix} M_S = -\frac{1}{2} \rightarrow \xi_- = \begin{pmatrix} 0 \\ 1 \end{pmatrix} \tag{1.6.1}$$

and the general wavefunctions as the combination of both spin states are given by

$$\Psi = (a\xi_+ + b\xi_-) \int c(E, \vec{p}\,) \exp\left[\frac{2\pi i}{h}(Et + \vec{p}\cdot\vec{r}\,)\right] \mathrm{d}E\mathrm{d}\vec{r} \tag{1.6.2}$$

The operator of the spin components in the x,y,z-directions are

$$\widetilde{S_q} = \frac{h}{4\pi}\sigma_q, \quad q = x, y, z$$

$$\sigma_x = \begin{pmatrix} 0 & 1 \\ 1 & 0 \end{pmatrix}, \sigma_y = \begin{pmatrix} 0 & -i \\ i & 0 \end{pmatrix}, \sigma_z = \begin{pmatrix} 1 & 0 \\ 0 & -1 \end{pmatrix}, \tag{1.6.3}$$

where σ_q are the Pauli matrices. For the Pauli matrices,

$$\sigma_x^2 = \sigma_y^2 = \sigma_z^2 = I, \quad I = \begin{pmatrix} 1 & 0 \\ 0 & 1 \end{pmatrix}$$

$$\sigma_x\sigma_y + \sigma_y\sigma_x = \sigma_x\sigma_z + \sigma_z\sigma_x = \sigma_y\sigma_z + \sigma_z\sigma_y = 0,$$

$$\sigma_x\sigma_y - \sigma_y\sigma_x = 2i\sigma_z,$$

$$\sigma_y\sigma_z - \sigma_z\sigma_y = 2i\sigma_x,$$

$$\sigma_z\sigma_x - \sigma_x\sigma_z = 2i\sigma_y \tag{1.6.4}$$

are satisfied. The square of the absolute value is given by

$$(\widetilde{S_x})^2 + (\widetilde{S_y})^2 + (\widetilde{S_z})^2 = \left(\frac{h}{2\pi}\right)^2 \frac{3}{4} I = \left(\frac{h}{2\pi}\right)^2 S(S+1)I, \tag{1.6.5}$$

which corresponds to the square of the absolute value of the orbital angular momentum is given by $\left(\frac{h}{2\pi}\right)^2 L(L+1)$ as shown by equation (1.5.38). Therefore, the electron spin satisfies all the relations with the angular momentum, except that the quantum numbers are given as half integers.

The spin is not derived from Schrödinger equation, because the spin wavefunction is a vector while the Hamiltonian is scalar.

1.6.2 Relativistic quantum mechanics (Klein–Gordon equation)

Based on the theory of special relativity, the relationship between energy and momentum is given by

$$E^2 = (mc^2)^2 + c^2 |\vec{p}|^2 \tag{1.6.6}$$

Using the energy and momentum operators shown in equation (1.5.13), the Klein–Gordon equation is derived as [24]

$$-\left(\frac{h}{2\pi}\right)^2 \frac{\partial^2}{\partial t^2} = -\left(\frac{h}{2\pi}\right)^2 c^2 \left[\frac{\partial^2}{\partial x^2} + \frac{\partial^2}{\partial y^2} + \frac{\partial^2}{\partial z^2}\right] + (mc^2)^2. \tag{1.6.7}$$

This equation derives the square of the energy (not energy itself), and the utility of the solutions is questionable. Negative probability and energy values are included in the solutions. The wavefunction is treated as a scalar; therefore, the effect of electron spin is not included. However, this equation is useful for treating Boson particles with zero spin (scalar particle), for which the wavefunction is treated as a scalar [25].

1.6.3 Dirac equation

Dirac proposed an equation for a Hamiltonian as a first-order differential equation as follows [26]:

$$\widetilde{H} = c\left[\widetilde{\alpha_x}\widetilde{p_x} + \widetilde{\alpha_y}\widetilde{p_y} + \widetilde{\alpha_z}\widetilde{p_z}\right] + \widetilde{\beta} mc^2 \tag{1.6.8}$$

with the requirement that $\widetilde{H}^2$ corresponds to the Klein–Gordon equation. For operators $\widetilde{\alpha_q}$ (q=x, y, z) and $\widetilde{\beta}$, the following relations must hold:

$$\widetilde{\alpha_q}^2 = \widetilde{\beta}^2 = 1$$

$$\widetilde{\alpha_x}\widetilde{\alpha_y} + \widetilde{\alpha_y}\widetilde{\alpha_x} = \widetilde{\alpha_y}\widetilde{\alpha_z} + \widetilde{\alpha_z}\widetilde{\alpha_y} = \widetilde{\alpha_z}\widetilde{\alpha_x} + \widetilde{\alpha_x}\widetilde{\alpha_z} = 0$$

$$\widetilde{\alpha_q}\widetilde{\beta} + \widetilde{\beta}\widetilde{\alpha_q} = 0. \tag{1.6.9}$$

Equation (1.6.9) cannot be satisfied for a scalar, but can be satisfied using 4×4 matrices

$$\alpha_q = \begin{pmatrix} 0 & \sigma_q \\ \sigma_q & 0 \end{pmatrix}, \beta = \begin{pmatrix} I & 0 \\ 0 & -I \end{pmatrix}, \tag{1.6.10}$$

where σ_q are the Pauli matrices (see equation (1.6.3)). The eigenfunctions are given by four-dimensional vectors

$$\vec{\Psi} = \begin{pmatrix} \vec{u} \\ \vec{w} \end{pmatrix}, \quad \vec{u} = \begin{pmatrix} u_1 \\ u_2 \end{pmatrix}, \quad \vec{w} = \begin{pmatrix} w_1 \\ w_2 \end{pmatrix}. \tag{1.6.11}$$

At first, we consider $\vec{p} = 0$ for simplicity. Then

$$\widetilde{H} = \begin{pmatrix} mc^2 & 0 & 0 & 0 \\ 0 & mc^2 & 0 & 0 \\ 0 & 0 & -mc^2 & 0 \\ 0 & 0 & 0 & -mc^2 \end{pmatrix} \tag{1.6.12}$$

and four eigenfunctions

$$\vec{\Psi} = \begin{pmatrix} 1 \\ 0 \\ 0 \\ 0 \end{pmatrix}, \begin{pmatrix} 0 \\ 1 \\ 0 \\ 0 \end{pmatrix}, \begin{pmatrix} 0 \\ 0 \\ 1 \\ 0 \end{pmatrix}, \begin{pmatrix} 0 \\ 0 \\ 0 \\ 1 \end{pmatrix} \tag{1.6.13}$$

are obtained. The energy eigenvalues are mc^2 for two solutions and $-mc^2$ for the other two solutions. Two solutions for each energy eigenvalue are interpreted as two spin states. The solution of negative rest energy made a chance to consider the existence of antiparticles as shown in section 3.1.

Here, we consider the solution of the Dirac equation in a free space as follows:

$$E\vec{u} = mc^2\vec{u} + c\left[\sigma_x\widetilde{p_x} + \sigma_y\widetilde{p_y} + \sigma_z\widetilde{p_z}\right]\vec{w}$$

$$E\vec{w} = -mc^2\vec{w} + c\left[\sigma_x\widetilde{p_x} + \sigma_y\widetilde{p_y} + \sigma_z\widetilde{p_z}\right]\vec{u}. \tag{1.6.14}$$

Taking $E' = E - mc^2$, equation (1.6.14) is rewritten as

$$E'\vec{u} = c\left[\sigma_x\widetilde{p_x} + \sigma_y\widetilde{p_y} + \sigma_z\widetilde{p_z}\right]\vec{w}$$

$$(2mc^2 + E')\vec{w} = c\left[\sigma_x\widetilde{p_x} + \sigma_y\widetilde{p_y} + \sigma_z\widetilde{p_z}\right]\vec{u} \tag{1.6.15}$$

With the classical approximation ($2mc^2 + E' \approx 2mc^2$)

$$\vec{w} = \frac{\left[\sigma_x \widetilde{p_x} + \sigma_y \widetilde{p_y} + \sigma_z \widetilde{p_z}\right]}{2mc}\vec{u}$$

$$E'\vec{u} = \frac{\widetilde{p_x}^2 + \widetilde{p_y}^2 + \widetilde{p_z}^2}{2m}\vec{u}, \tag{1.6.16}$$

which corresponds to the Schrödinger equation.

When there is electromagnetic field (electric field $\vec{E} = -\nabla\cdot\Phi_{el} - \partial\vec{A}/\partial t$ and magnetic field $\vec{B} = \nabla \times \vec{A}$), we use the exchange $E \to E - q_e\Phi_{el}$ and $\vec{p} \to \vec{p} + q_e\vec{A}$ taking $\vec{A} = \frac{1}{2}(-yB, xB, 0)$ and $\vec{B} = (0,0, B)$, where q_e is the electric charge. The product between $\sigma_{x,\,y}\left(p_{x,\,y} + q_e A_{x,\,y}\right)$ is given by

$$\left[\sigma_x\left(\frac{h}{2\pi i}\frac{\partial}{\partial x} - \frac{q_e yB}{2}\right)\right]^2 = \begin{pmatrix} -\left(\frac{h}{2\pi}\right)^2\frac{\partial^2}{\partial x^2} - q_eBy\left(\frac{h}{2\pi i}\right)\frac{\partial}{\partial x} + \left(\frac{q_e yB}{2}\right)^2 & 0 \\ 0 & -\left(\frac{h}{2\pi}\right)^2\frac{\partial^2}{\partial x^2} - q_eBy\left(\frac{h}{2\pi i}\right)\frac{\partial}{\partial x} + \left(\frac{q_e yB}{2}\right)^2 \end{pmatrix}$$

$$\left[\sigma_y\left(\frac{h}{2\pi i}\frac{\partial}{\partial y} + \frac{q_e xB}{2}\right)\right]^2 = \begin{pmatrix} -\left(\frac{h}{2\pi}\right)^2\frac{\partial^2}{\partial y^2} + q_eBx\left(\frac{h}{2\pi i}\right)\frac{\partial}{\partial y} + \left(\frac{q_e xB}{2}\right)^2 & 0 \\ 0 & -\left(\frac{h}{2\pi}\right)^2\frac{\partial^2}{\partial y^2} + q_eBx\left(\frac{h}{2\pi i}\right)\frac{\partial}{\partial y} + \left(\frac{q_e xB}{2}\right)^2 \end{pmatrix}$$

$$\sigma_x\left(\frac{h}{2\pi i}\frac{\partial}{\partial x} - \frac{q_e yB}{2}\right)\sigma_y\left(\frac{h}{2\pi i}\frac{\partial}{\partial y} + \frac{q_e xB}{2}\right) + \sigma_y\left(\frac{h}{2\pi i}\frac{\partial}{\partial y} + \frac{q_e xB}{2}\right)\sigma_x\left(\frac{h}{2\pi i}\frac{\partial}{\partial x} - \frac{q_e yB}{2}\right) = \frac{q_e h}{2\pi}B\begin{pmatrix} 1 & 0 \\ 0 & -1 \end{pmatrix}. \tag{1.6.17}$$

Taking the electron orbital angular momentum and the spin as

$$L_z = \left[x\left(\frac{h}{2\pi i}\frac{\partial}{\partial y}\right) - y\left(\frac{h}{2\pi i}\frac{\partial}{\partial x}\right)\right]$$

$$S_z = \frac{h}{4\pi}\sigma_z, \tag{1.6.18}$$

the energy of the electron is given by

$$E'\vec{u} = \left[\frac{\widetilde{p_x}^2 + \widetilde{p_y}^2 + \widetilde{p_z}^2}{2m} + \frac{q_e}{2m}L_zB + q_e\Phi_{el}\right]I\vec{u} + 2\frac{q_e}{2m}BS_z\vec{u}. \tag{1.6.19}$$

For an electron, the Zeeman energy shift E_Z is often described by

$$L_z = \frac{h}{2\pi}M_L$$

$$S_z = \frac{h}{2\pi}M_S$$

$$E_Z = \mu_B[M_L + g_S M_S]B$$

$$\mu_B = \frac{eh}{4\pi m_e}. \tag{1.6.20}$$

The g-factor of the electron spin g_S is obtained as exactly two from equations (1.6.19)–(1.6.20), but it was experimentally determined to be 2.0023 [27]. This discrepancy is the 'anomalous magnetic moment', which is derived based on the quantum electrodynamics [28] taking the energy fluctuations in a vacuum into account (section 3.13). The existence of the Zeeman energy shift induced by the electron spin is not a relativistic effect, but it was obtained from the Dirac equation, which requires the Hamiltonian be treated as matrices.

Relativistic effects are derived by performing a more detailed analysis from equation (1.6.15) assuming $\vec{A} = 0$ and $\Phi_{el} \neq 0$ as follows.

$$\begin{gathered}
(2mc^2 + E')\vec{w} = c\left[\sigma_x\tilde{p}_x + \sigma_y\tilde{p}_y + \sigma_z\tilde{p}_z\right]\vec{u} \\
\vec{w} = \frac{c\left[\sigma_x\tilde{p}_x + \sigma_y\tilde{p}_y + \sigma_z\tilde{p}_z\right]\vec{u}}{2mc^2 + E'} = \frac{c\left[\sigma_x\tilde{p}_x + \sigma_y\tilde{p}_y + \sigma_z\tilde{p}_z\right]\vec{u}}{2mc^2} - \frac{E'c\left[\sigma_x\tilde{p}_x + \sigma_y\tilde{p}_y + \sigma_z\tilde{p}_z\right]\vec{u}}{(2mc^2)^2} \\
E'\vec{u} = E'_{cl}\vec{u} + E'_{\mathrm{rel}}\vec{u} \\
E'_{cl} = \frac{\tilde{p}_x^{\,2} + \tilde{p}_y^{\,2} + \tilde{p}_z^{\,2}}{2m} + q_e\Phi_{el} \\
E'_{\mathrm{rel}} = -c\left[\sigma_x\tilde{p}_x + \sigma_y\tilde{p}_y + \sigma_z\tilde{p}_z\right]\frac{E'c\left[\sigma_x\tilde{p}_x + \sigma_y\tilde{p}_y + \sigma_z\tilde{p}_z\right]}{(2mc^2)^2}.
\end{gathered} \tag{1.6.21}$$

To calculate the relativistic correction, E'_{rel} was calculated by taking $E' = E'_{cl}$ for simplicity. Then E'_{rel} is separated into

$$E'_{\mathrm{rel}} = K_{\mathrm{rel}} + P_{rel}. \tag{1.6.22}$$

K_{rel} is the relativistic correction of the kinetic energy, which is given by

$$K_{\mathrm{rel}} = -c\left[\sigma_x\tilde{p}_x + \sigma_y\tilde{p}_y + \sigma_z\tilde{p}_z\right]\frac{\tilde{p}_x^{\,2} + \tilde{p}_y^{\,2} + \tilde{p}_z^{\,2}}{2m}\frac{c\left[\sigma_x\tilde{p}_x + \sigma_y\tilde{p}_y + \sigma_z\tilde{p}_z\right]\vec{u}}{\left(2mc^2\right)^2} = -\frac{\tilde{p}_x^{\,4} + \tilde{p}_y^{\,4} + \tilde{p}_z^{\,4}}{8m^3c^2}. \tag{1.6.23}$$

This term is derived as the third term of

$$E = mc^2\sqrt{1 + \left(\frac{|\vec{p}\,|}{mc}\right)^2} = mc^2 + \frac{|\vec{p}\,|^2}{2m} - \frac{|\vec{p}\,|^4}{8m^3c^2}. \tag{1.6.24}$$

The relativistic effect induced by Φ_{el} is given by:

$$P_{\mathrm{rel}} = -\left[\sigma_x\tilde{p}_x + \sigma_y\tilde{p}_y + \sigma_z\tilde{p}_z\right]q_e\Phi_{el}\frac{\left[\sigma_x\tilde{p}_x + \sigma_y\tilde{p}_y + \sigma_z\tilde{p}_z\right]}{4m^2c^2}. \tag{1.6.25}$$

Assuming that Φ_{el} is spherically symmetric,

$$P_{rel} = P_d + P_{fs}$$

$$P_d = \frac{q_e}{4m^2c^2}\left(\frac{h}{2\pi}\right)^2 \frac{\partial \Phi_{el}}{r\partial r}\begin{pmatrix} x\frac{\partial}{\partial x} + y\frac{\partial}{\partial y} + z\frac{\partial}{\partial z} & 0 \\ 0 & x\frac{\partial}{\partial x} + y\frac{\partial}{\partial y} + z\frac{\partial}{\partial z} \end{pmatrix} = \frac{q_e}{4m^2c^2}\left(\frac{h}{2\pi}\right)^2 \frac{\partial^2 \Phi_{el}}{\partial r^2} I$$

$$P_{fs} = \frac{q_e}{4m^2c^2}\frac{h}{2\pi i}\frac{\partial \Phi_{el}}{r\partial r}\left[L_x\sigma_x + L_y\sigma_y + L_z\sigma_z\right] = \frac{q_e}{2m^2c^2}\frac{\partial \Phi_{el}}{r\partial r}\vec{L}\cdot\vec{S}. \tag{1.6.26}$$

Here, P_d is the Darwin's term and P_{fs} is the fine structure term, given as the coupling between the orbital angular momentum and the spin. Using the sum angular momentum $\vec{J} = \vec{L} + \vec{S}$,

$$\vec{L}\cdot\vec{S} = \frac{J(J+1) - L(L+1) - S(S+1)}{2}. \tag{1.6.27}$$

The $(J,\ M_J)$ state is given by the coupling of the $(L,\ M_L)$ and $(S,\ M_S)$ states satisfying

$$M_J = M_L + M_S. \tag{1.6.28}$$

For an electron ($S = 1/2$), the $J = L + 1/2$ and $J = L - 1/2$ states denote the states in which the orbital angular momentum and spin are parallel and antiparallel, respectively. Here, we consider the Zeeman shift in the $(J,\ M_J)$ state with the $(L = 1,\ M_L = 0, \pm 1)$ and $(S = 1/2,\ M_S = \pm 1/2)$ states. Each state with zero-magnetic field is given by

$$\left(J = \frac{3}{2},\ M_J = \pm\frac{3}{2}\right):\ (M_S,\ M_L) = \left(\pm\frac{1}{2},\ \pm 1\right)$$

$$\left(J = \frac{3}{2},\ M_J = \pm\frac{1}{2}\right):\ (M_S,\ M_L) = \sqrt{\frac{2}{3}}\left(\pm\frac{1}{2},\ 0\right) + \sqrt{\frac{1}{3}}\left(\mp\frac{1}{2},\ \pm 1\right)$$

$$\left(J = \frac{1}{2},\ M_J = \pm\frac{1}{2}\right):\ (M_S,\ M_L) = \sqrt{\frac{1}{3}}\left(\pm\frac{1}{2},\ 0\right) - \sqrt{\frac{2}{3}}\left(\mp\frac{1}{2},\ \pm 1\right). \tag{1.6.29}$$

When there is a magnetic field, the $(J = 3/2,\ M_J = \pm 1/2)$ and $(J = 1/2,\ M_J = \pm 1/2)$ states are mixed. Using equations (1.5.48) and (1.6.20), the Zeeman shift induced by the magnetic field in the z-direction B at each state is given by

$\left(J = \frac{3}{2},\ M_J = \pm\frac{3}{2}\right)$: $\pm\mu_B\left[1 + \frac{g_S}{2}\right]B$

$\left(J = \frac{3}{2},\ M_J = \pm\frac{1}{2}\right)$: $\pm\mu_B\left[\frac{1}{3} + \frac{g_S}{6}\right]B + \frac{2}{9}\frac{\left[\mu_B(g_S - 1)B\right]^2}{h\Delta_{fs}}$ (low magnetic field)

:$\pm\mu_B\frac{g_S}{2}B$ (high magnetic field)

$\left(J = \frac{1}{2},\ M_J = \pm\frac{1}{2}\right)$: $\pm\mu_B\left[\frac{2}{3} - \frac{g_S}{6}\right]B - \frac{2}{9}\frac{\left[\mu_B(g_S - 1)B\right]^2}{h\Delta_{fs}}$ (low magnetic field)

:$\pm\mu_B\left(1 - \frac{g_S}{2}\right)B$ (high magnetic field)

$$\Delta_{fs}\text{: transition frequency between the } J = \frac{3}{2} \text{ and } J = \frac{1}{2} \text{ states.} \tag{1.6.30}$$

The energy eigenvalue of the electron in a hydrogen atom or hydrogen like ions (one electron and the nucleus with a charge of + *Ze*) is obtained strictly from the Dirac equation as

$$E_{n,J} = \frac{m_e c^2}{\sqrt{1 + \left(\frac{Z\alpha}{n - J - \frac{1}{2} + \sqrt{\left(J + \frac{1}{2}\right)^2 - (Z\alpha)^2}}\right)^2}}$$

$$\alpha = \frac{e^2}{2\varepsilon_0 hc} \quad \text{(fine structure constant).} \tag{1.6.31}$$

For the value of $\alpha = 0.00\ 116$, the following approximation is valid for a small Z.

$$\sqrt{\left(J + \frac{1}{2}\right)^2 - (Z\alpha)^2} = \left(J + \frac{1}{2}\right) - \frac{(Z\alpha)^2}{2\left(J + \frac{1}{2}\right)}$$

$$E_{n,J} = \frac{m_e c^2}{\sqrt{1 + \left(\frac{Z\alpha}{n - \frac{(Z\alpha)^2}{2\left(J + \frac{1}{2}\right)}}\right)^2}} = m_e c^2 - \frac{m_e c^2 Z^2 \alpha^2}{2\left(n - \frac{(Z\alpha)^2}{2\left(J + \frac{1}{2}\right)}\right)^2}$$

$$\text{taking } \frac{1}{\left(n - \frac{(Z\alpha)^2}{2\left(J + \frac{1}{2}\right)}\right)^2} = \frac{1}{n^2} + \frac{(Z\alpha)^2}{n^3\left(J + \frac{1}{2}\right)}$$

$$= m_e c^2 - \frac{m_e Z^2 e^4}{8\varepsilon_0^2 h^2 n^2} - \frac{m_e Z^4 e^4}{8\varepsilon_0^2 h^2 n^3} \frac{\alpha^2}{\left(J + \frac{1}{2}\right)}. \tag{1.6.32}$$

The first term denotes the rest energy of an electron, and the second term represents the energy eigenvalue obtained using the Schrödinger equation. The third term denotes the relativistic effect depending on J. Note that the relative motion between the electron and nucleus, and the motion of the center of mass are not separable with the theory of relativity. Therefore, it is often treated by changing m_e to μ_e (reduced mass) in the second and third terms of equation (1.6.32). The dependence of the energy eigenvalues on α is significant for highly charged ions.

The Dirac equation using a 4×4 matrix is applicable for the particles with $S = 1/2$. For the particles with $S = 0$, the wavefunction is scalar, and the Klein–Gordon equation is applicable [24]. For the particles with other spin values, the following equations are applicable.

The Rarita–Schwiner equation massive particles with $S = 3/2$ [29].

The Bargmann–Wigner equation free particles with arbitral spin [30].

The Bargmann–Wigner equation is a set of linear partial differential equations with $2S$ components. The calculation of each component was based on the Dirac equation ($2S = 1$). However, it is difficult to use the Bargmann–Wigner equation for particles in an electromagnetic field.

References

[1] Milham W I 1923 *Time & Timekeepers, Including the History, Construction, Care, and Accuracy of Clocks and Watches* (New York: MacMillan) pp 1–88
[2] Filonovich S R 1986 *The Greatest Speed* (Moscow: Mir Publishers) p 285
[3] Bradley J 1728 *Philos. Trans. R. Soc. London* **35** 637–61
[4] Fizeau H 1849 *Comptes Rendus Hebdomadaires des Seances de l'Academie des Science* **29** 90
[5] Foucault L 1862 *Comptes Rendus Hebdomadaires des Seances de l'Academie des Science* **55** 501
[6] Huray P G 2010 *Maxwell's Equations* (Wiley-IEEE) p 22
[7] Michelson A A 1881 *Am. J. Sci.* **22** 120
[8] Brillet A and Hall J L 1979 *Phys. Rev. Lett.* **42** 549
[9] London R 1973 *The Quantum Theory of Light* 3rd edn (Cambridge University)
[10] Einstein A 1916 *Relativity-The Special and General Theory* translated 1920 (New York: H. Holt and Company)
[11] Kaminer I *et al* 2015 *Nat. Phys.* **11** 261
[12] Dyson F W *et al* 1920 *Philos. Trans. R. Soc. London* **220A** 291
[13] Thomson J J 1897 Cathode rays *Phil. Mag.* **44** 293
[14] Rutherford E 1911 *Phil. Mag. Ser.* **6** 21 https://lawebdefisica.com/arts/structureatom.pdf
[15] Balmer J J 1885 *Ann. Phys.* **261** 80 (in German)
[16] Lyman T 1906 *New Ser.* **13** 125
[17] Bohr N 1913 *Phil. Mag. Ser. 6* **26** 1
[18] de Broglie L V 1925 *Ann. Phys.* **10** 22
[19] Ozawa M 2003 *Phys. Rev.* A **67** 042105
[20] Erhart J *et al* 2012 *Nat. Phys. (N. Y.)* **8** 185
[21] Shimizu F, Shimizu K and Takuma H 1992 *Jpn. J. Appl. Phys.* **31** L436
[22] Gerlach W and Stern O 1922 *Z. Phys.* **9** 349
[23] Liboff R L 2002 *Introductory Quantum Mechanics* (Addison-Wesley/Pauli matrices – Wikipedia)
[24] Klein O 1926 *Z. Phys.* **37** 895
[25] Pauli W and Weisskopf V 1934 *Helv. Phys. Acta* **7** 709
[26] Dirac P A M 1942 Bakerian Lecture. *Proc. R. Soc. A: Math. Phys. Eng. Sci.* **180** 1–39
[27] Kusch R and Foley H M 1948 *Phys. Rev.* **74** 250
[28] Schwinger J 1948 *Phys. Rev.* **73** 416
[29] Rarita W and Schwiner J 1941 *Phys. Rev.* **60** 61
[30] Bargmann V and Wigner E P 1948 *Proc. Natl Acad. Sci. USA* **34** 211–23
[31] Kajita M 2018 *Measuring Time* (Bristol: IOP Publishing)
[32] Kajita M 2020 *Cold Atoms and Molecules* (Bristol: IOP Publishing)

Chapter 2

Recently solved mysteries

This chapter introduces experimental procedures in modern physics. Development of the atomic clock made it possible to confirm the relativistic effects in progress of time. The invention of the laser contributed significantly to the development of modern physics; optical pumping to a single atomic state, laser spectroscopy of atoms and molecules, laser cooling of atoms and molecules. Laser cooling was a significant contributor to the confirmation of quantum mechanics by Bose–Einstein condensation, the entangled state between multiple particles, and the 'Schrödinger cat' phenomenon. The measurement uncertainty of time and frequency was drastically reduced by measuring the optical transition frequencies of laser cooled atoms. The quantum effects have also contributed to real life; inertial navigation using atomic interferometry, quantum computer, atomic magnetometer etc. The gravitational wave was detected using a laser interferometer.

2.1 Development of the atomic clock

As shown in figure 2.1, It has been clarified that atoms and molecules absorb or emit electromagnetic waves (light, microwave) only when its frequency ν corresponds to the transition frequencies ν_0 satisfying

$$\nu_0 = \frac{\Delta E}{h}, \tag{2.1.1}$$

where ΔE is the difference in energy between different atomic or molecular quantum states. The transition frequencies can be measured with an uncertainty much lower than any other physical values because the transition is suppressed significantly with a slight shift of ν from ν_0. Therefore, atomic transition frequencies seemed to be a good standard for time and frequency (molecular transition rates are lower than atomic transitions).

In 1956, an atomic clock based on the transition frequency of cesium (Cs) atom in the microwave region (transition between hyperfine states 'a' and 'b' given by the

doi:10.1088/978-0-7503-6239-9ch2

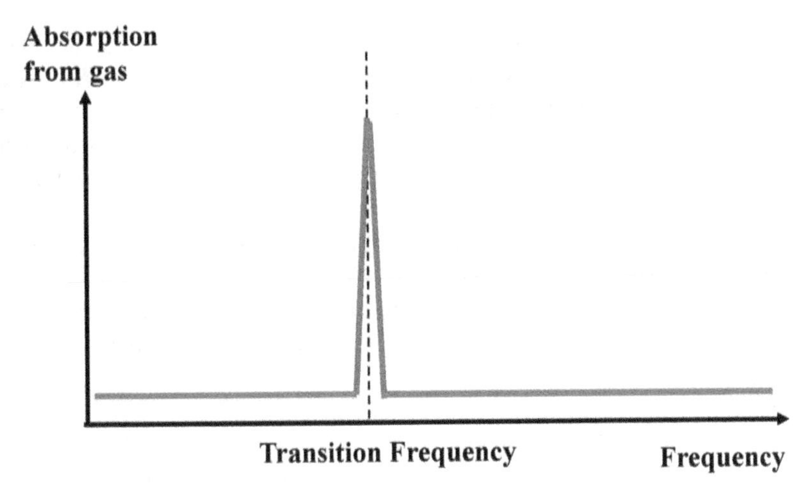

Figure 2.1. The electromagnetic wave (light or microwave) is absorbed by gaseous atoms or molecules with a specific frequency ν which satisfies equation (2.1.1).

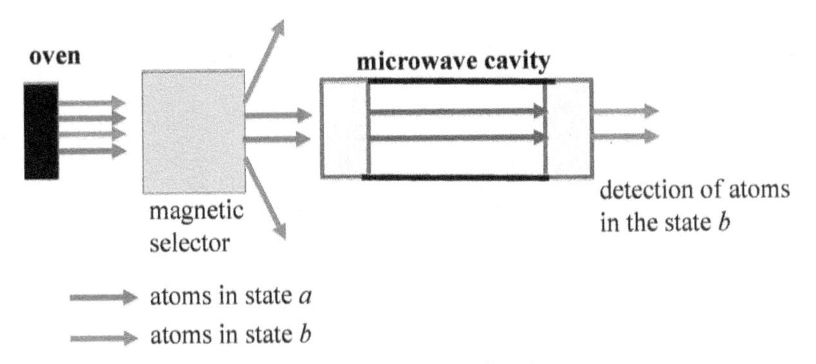

Figure 2.2. The structure of cesium atomic clock using thermal atomic beam and magnetic state selector.

interaction between the electron spin and nuclear spin) was constructed [1]. An atomic beam from an oven passes through a magnetic selector, with which atoms in the 'a' state are focused while atoms in the 'b' state are deflected. An atomic beam localized in the 'a' state passes through a microwave cavity. If the microwave frequency is resonant to the a → b transition, atoms in the 'b' state are detected, as shown in figure 2.2. The microwave frequency is stabilized to the transition frequency giving feedback to maximize the a → b transition rate. The fractional uncertainty was at first 10^{-10}, and it has been reduced to 10^{-13} also for a commercial type. In 1967, the Cs transition frequency (9 192 631 770 Hz) was defined to be the standard of the time and frequency [2]. Atomic clocks are more advantageous than previous clocks (for example, crystal clocks) to reduce the measurement uncertainties for the following reasons:

(i) The atomic transition frequency is discrete, and the transition rate is reduced drastically with a slight shift in the frequency of the electromagnetic wave. The previous clocks are based on the oscillation frequency which can take continuous values (for example, crystal oscillation).

(ii) In a gaseous state, the atoms are isolated from one another. The characteristics of a single atom are therefore uniform all over the world. In a solid material, the bonding state between neighboring atoms cannot be perfectly the same also with the same trademark because some impurity is always included. The operation of clocks using solids fluctuate by the change of circumstances (temperature, humidity etc).

(iii) The atomic energy structure is determined mostly by the Coulomb force between the nucleus and the electrons, which is uniform for all atoms.

(iv) The atomic transition frequencies are much higher than the oscillation frequency of quartz crystals, and the time is measured with finer scale division.

But (i)–(iii) are not perfectly satisfied even with the atomic clocks. Appendix C indicates the statistical and systematic uncertainties of the measured transition frequencies. The measurement uncertainty was drastically reduced after the development of lasers (section 2.8).

2.2 Confirmation of relativistic effects with time

As shown in section 1.3, theory of special relativity shows that the time goes slower in a moving frame (figure 2.3). Transition frequency of atoms or molecules moving with velocity v is observed as a lower value than the real value with a factor of $\sqrt{1-(v/c)^2}$, which is called the 'quadratic Doppler effect'. The ratio of the

Figure 2.3. From the theory of special relativity, time goes slower in a moving frame.

Figure 2.4. From the theory of general relativity, time goes slower with the gravitational potential.

fractional shift with this effect is 5×10^{-13} with the velocity of 300 m s^{-1}, which is difficult to detect with crystal clocks.

There were frequency discrepancies between different Cs atomic clocks using thermal atomic beams [3]. The quadratic Doppler effect can be estimated by measuring the atomic velocity distribution, which depends on the structure of the magnetic selector. After giving the correction of the quadratic Doppler effect, the frequency discrepancies were drastically reduced.

The theory of general relativity indicates (section 1.4) that the time goes slower with the gravitational potential of $m\Phi_G$(m: mass, $\Phi_G < 0$) with a factor of $\sqrt{1 + (2\Phi_G/c^2)}$ (figure 2.4).

Therefore, the observed atomic transition frequency is lower in a gravitational potential (gravitational red shift). The measured transition frequency becomes higher with the ratio of 10^{-16} with higher altitude of 1 m. This effect was measured by comparing the frequencies of a Cs atomic clock when located in a valley and on a nearby mountain [4].

To obtain the measurement uncertainties below 10^{-14} with the transition frequencies of atoms or molecules, the relativistic effects should be estimated and corrected.

2.3 Development of lasers

Since the first laser was demonstrated in 1960, many kinds of lasers have been developed and they contribute a great deeal to the development of new research fields; observation of the spectrum of atoms and molecules, measurement of the frequencies of light, and laser cooling of atoms and molecules.

The word 'LASER' is the acronym for '**L**ight **A**mplifier by **S**timulated **E**mission of **R**adiation'. The transition between two energy states 1 and 2, having energies of E_1 and E_2 ($E_1 < E_2$) and the populations $\rho(1)$ and $\rho(2)$. For simplicity, the degeneracies at both states are assumed to be 1. With thermal equilibrium state

(Appendix A), $\rho(1) > \rho(2)$. By the irradiation of a light with the transition frequency, the $1 \rightarrow 2$ transition is induced with the rate of $B_t I(\rho(1) - \rho(2))$, where B_t is the coefficient, and I is the intensity of the incident light wave. The incident light wave is damped with this transition, so that the total energy is conserved. There is a $2 \rightarrow 1$ transition emitting a fluorescence light with the rate of $A_t\rho(2)$, which is called the spontaneous emission transition. The phase and the direction of the emitted fluorescence light is totally random. The equilibrium population is given by $B_t I(\rho(1) - \rho(2)) = A_t\rho(2)$. When the transition is induced by the blackbody radiation (section 1.2.3), $\rho(2)/\rho(1) = \exp[-(E_2 - E_1)/k_B T]$ as shown in Appendix A (T: thermal dynamic temperature).

When $\rho(1) < \rho(2)$ (called the inversion population), the $2 \rightarrow 1$ induced transition is more dominant than the $1 \rightarrow 2$ transition. The incident light wave is amplified with the same phase and the propagation direction, as shown in figure 2.5. Pumping to the energy state 2 (some artificial treatment to increase $\rho(2)$) is required to obtain the inversion population between the 1 and 2 states. It is realistic to attain the inversion population with the three states system. We consider three energy states 0, 1, and 2 with energies of E_0, E_1, and E_2 ($E_0 < E_1 < E_2$) and populations of $\rho(0)$, $\rho(1)$, and $\rho(2)$. When $\rho(0) \gg \rho(1)$, the inversion population between 1 and 2 states is obtained by the $0 \rightarrow 2$ pumping with the realistic rate.

The amplification of light by laser medium is generally negligibly small except for when the laser medium is contained in a cavity, as shown in figure 2.5 (the Fabry–Perot cavity is considered in this figure). When the amplification (gain) of the light given by the laser medium is higher than the loss inside the cavity, the light with the resonant mode to the cavity repeat amplification while reflecting in the cavity. As the light intensity is increased, $\rho(2)$–$\rho(1)$ becomes smaller by the higher stimulated emission rate (saturation effect). Then the gain becomes smaller, and the amplification stops with the light intensity where the gain balances with the cavity loss. The

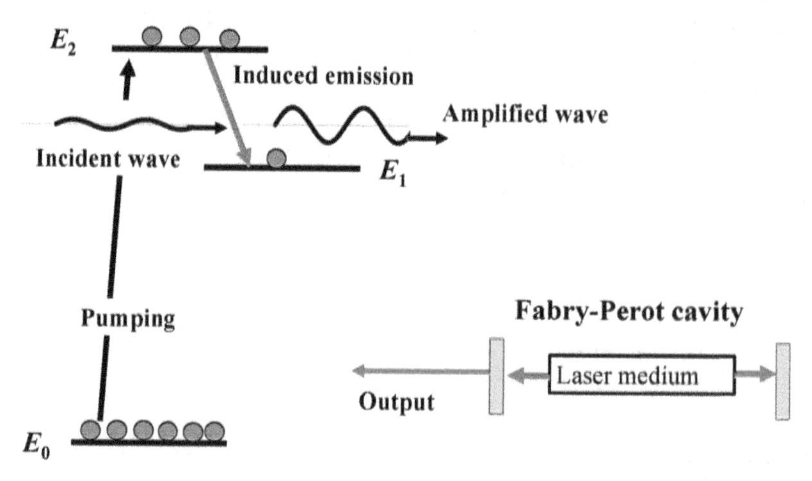

Figure 2.5. Incident wave is amplified by stimulated emission when inversion population between the 1 and 2 states is formed by the 0→2 pumping. Laser oscillator using Fabry–Perot cavity containing a laser medium is also indicated.

wavelength of the light resonant to the Fabry–Perot cavity with the length of L is $2L/n_m$, where n_m is the integer, therefore, the resonance frequency is $n_m c/2L$. The linewidth of each frequency component is $c/2LN_r$ (N_r: mean reflection time), which converges to zero with N_r, $L \rightarrow \infty$. When the laser oscillation is possible only with the single cavity resonance frequency, the laser oscillates with a single frequency. To select a single resonance frequency, a grating mirror (high reflectivity only with a limited frequency area) is often used.

The special properties of light obtained by the laser oscillation are:

(i) uniform phase, therefore, interference is easily observed;
(ii) propagation to a single direction parallel to the cavity;
(iii) narrow spectrum linewidth at each frequency component;
(iv) ultra-short pulse is obtained from the interference between many frequency components;
(v) focusing to the wavelength size is possible.

A laser is an instrument with which we can manipulate the parameters (frequency etc) with a high accuracy, and it provided a revolution in modern experimental physics.

Frequency transform of laser lights are performed to obtain light sources with frequencies, which are difficult to obtain from the direct laser oscillation. The frequency of laser light can be integer multiplied by the interaction with a crystal. Frequency doubled laser light is particularly often used as the light source in the ultra-violet region. Irradiating two laser lights with the frequencies of $\nu_{1,2}$ to a crystal, light with the frequency of $|\nu_1 \pm \nu_2|$ is obtained. The light source in the infrared or far-infrared region is often obtained from the frequency difference of two laser lights.

2.4 Optical pumping using laser light

We can manipulate the quantum state of atoms or molecules using laser lights. The population in the different quantum states in atoms or molecules (neutral or ion) are of the same order when the energy difference between them is smaller than $k_B T$ as shown in Appendix A (k_B: Boltzmann constant, T: thermal dynamic temperature). Optical pumping is a method to localize them in a single state. Here we consider with two ground states $g_{1,2}$ and one excited state e. When the $g_2 \rightarrow e$ transition is excited by a laser light, the deexcitation is caused by the spontaneous emission transitions to the g_1 or g_2 state. The $g_2 \rightarrow e \rightarrow g_1$ transition is caused with this procedure, while no transition is induced from the g_1 state. The selective excitation from the g_2 state is possible because of the narrow frequency bandwidth of laser light. The population is finally totally localized to the g_1 state, as shown in figure 2.6. This procedure is called 'optical pumping'. No fluorescence is observed after the optical pumping. Optical pumping has been used for example with the Cs atomic clock to localize atoms in one of two hyperfine states before interacting with microwave [5].

Optical pumping is a technology to reduce the entropy (parameter of randomness) of atoms or molecules, as shown in Appendix A. However, the increase of the

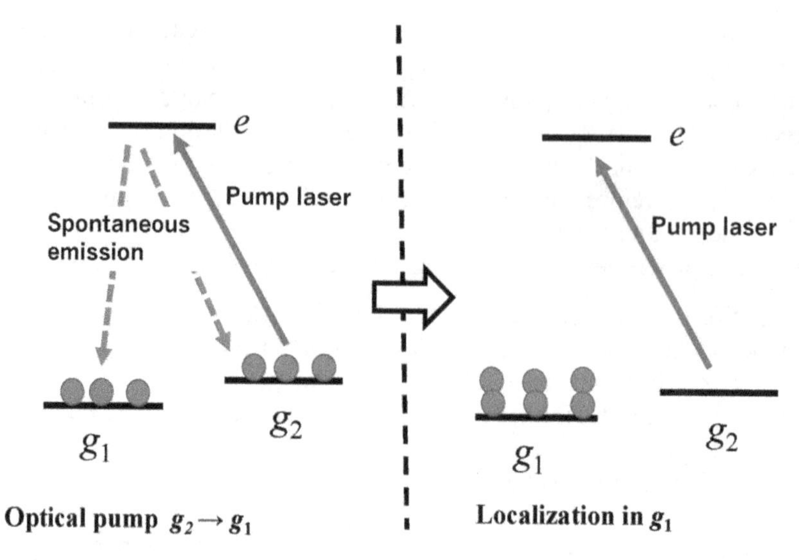

Figure 2.6. The procedure of optical pumping with the energy structure of $g_{1,2}$ and e states. With a laser light, which is resonant to the g_2–e transition, the distribution is localized to the g_1 state. Reproduced from [19].

entropy is obtained by transforming the laser light with a uniform phase to a fluorescence light with a random phase. Therefore, the principle of 'increase of entropy' is not violated.

2.5 Laser spectroscopy and atomic clocks using transitions in the optical region

The measurement of the transition frequencies of atoms or molecules is useful to search for their quantum energy structures. The frequency bandwidth of laser light is narrow, and transition frequencies are measured with high accuracy observing the transition signal with a high signal-to-noise ratio. The control of the optical path is easy when laser light is used because of its uniform propagation direction.

Since 1970, the spectroscopic research into polar molecules has been performed using CO_2 or N_2O lasers with many discrete oscillation frequencies. Their spectra were observed by inducing the shift of the transition frequencies by applying a DC electric field (laser Stark spectroscopy), because frequency scanning is not possible with these lasers. The spectrum was at first observed with NH_3 molecules, having simple energy structure and high absorption coefficient [6]. This method has also been applied for other molecules (H_2CO [7], CH_3F [8] etc). Frequency tunable dye lasers and diode lasers were later developed and the observation of spectrum by frequency scanning became possible [9]. Atomic or molecular spectra are also used for the stabilization of laser frequency by giving feedback to maximize the transition rate.

The development of laser spectroscopy provided the opportunity to also develop atomic clocks based on the transition frequencies in the optical region. In principle, optical transition frequencies are more advantageous than microwave transition frequencies to get lower measurement uncertainties, because the period of one second is measured with a finer scale. However, direct measurement of the frequencies of laser light was not realistic, while the microwave frequency was measured using frequency counters. Until the frequency comb was developed, the laser frequency was measured using wavelength meters with an uncertainty of 10^{-7}.

At the beginning of the 21st century, the frequency comb system was developed [10], which made it possible to measure the frequency of laser light with ultra-low measurement uncertainty. As shown in section 2.3, the resonance frequencies of the cavity with a length of L is $n_m c/2L$, where n_m is the integer. When the laser oscillation is possible for the frequency components of $N_{\min} \leqslant n_m \leqslant N_{\max}$ with a uniform phase, the laser light is a pulse laser with a repetition rate of $f_{\text{rep}} = c/2L$ (inverse of the period of the round trip in the cavity of the light) as shown in figure 2.7. The control of cavity loss is performed so that the laser oscillation is possible only with the uniform phase for all frequency components (mode locking). The laser material has a refraction rate with a slight dependence on the frequency and the effective optical length of cavity is simply expressed as $L_{\text{eff}} = \frac{L}{\left(1 + \frac{\delta_L}{n_m c}\right)}$. The actual frequency components are given by $\nu_{n_m} = n_m f_{\text{rep}} + f_{\text{ceo}} \left(f_{\text{ceo}} = f_{\text{rep}} \delta_L\right)$, which can be determined measuring f_{rep} and f_{ceo} using a frequency counter. Here, f_{ceo} is obtained from the beat signal between frequency doubled ν_{n_m} component (section 2.3) and ν_{2n_m} component ($f_{\text{ceo}} = 2\nu_{n_m} - \nu_{2n_m}$). The frequency of arbitral laser light ν_L is measured from the minimum beat frequency $f_{\text{beat}} = \left| \nu_{n_m} - \nu_L \right|$, after determining n_m using the wavelength meter. Development of optical atomic clocks became one of the hottest research areas after the development of the frequency comb.

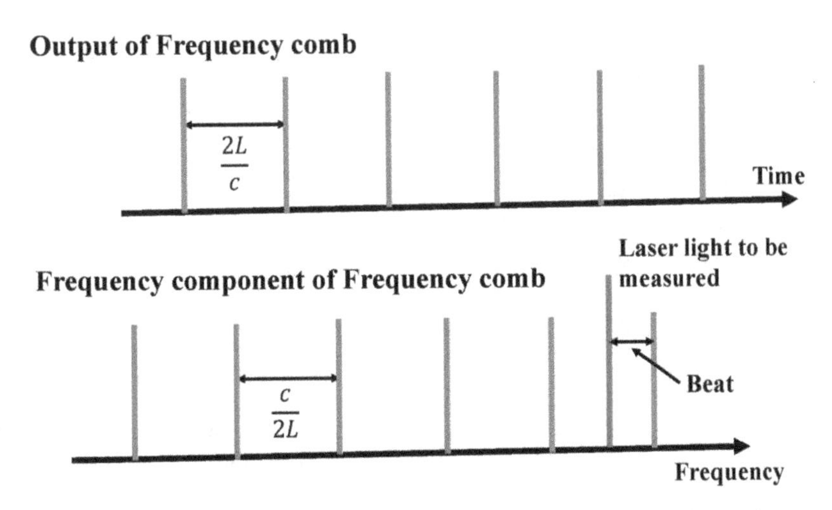

Figure 2.7. Output as pulse and the frequency spectrum of the frequency comb.

2.6 Laser cooling

Another revolution for atomic physics was due to lasers. The development of laser cooling technology made it possible to decelerate gaseous atoms from the speed of 200–400 m s^{-1} to slower than 10 cm s^{-1}. Here we assume an atom with one ground state and one excited state. When atoms absorb laser light (energy and momentum of photon), atoms are transformed to an excited energy state and simultaneously experience a force along the direction of light propagation, as shown in figure 2.8 [11]. Afterwards, atoms emit photons in random directions (spontaneous emission) and return to the ground state. With the emission of photons, atoms receive a recoil force; nevertheless, it is randomly directed and hence averages to zero. Repeating the cycle of absorption and spontaneous emission, atoms experience a net force along the direction of light propagation.

To cool atoms of random velocities, two laser beams of frequency lower than the transition frequency (red detuned) irradiate the atoms from opposing directions. Because of the first-order Doppler effect, atoms interact only with opposing light and all atoms are decelerated as shown in figure 2.8. Note that the spontaneous emission transition with the rate of Γ makes a spectrum broadening of $\Gamma/2\pi$ as shown in Appendix C. As the atoms decelerate and the Doppler shift becomes smaller than $\Gamma/2\pi$, the difference in the interaction forces from the two opposing laser beams becomes smaller. Cooling stops when the deceleration force becomes negligible small. The attainable kinetic energy temperature (called the Doppler limit) is given by $T_{\text{Doppler}} = h\Gamma/4\pi k_B$. Using the transition with large Γ, the cooling force is strong because of the rapid repetition of the cooling transition cycle. But transitions with small Γ are advantageous to get lower kinetic energy. Two step cooling is sometimes performed: first cooling using transitions with large Γ for rapid cooling, and the second cooling using transitions with small Γ to attain lower kinetic energy.

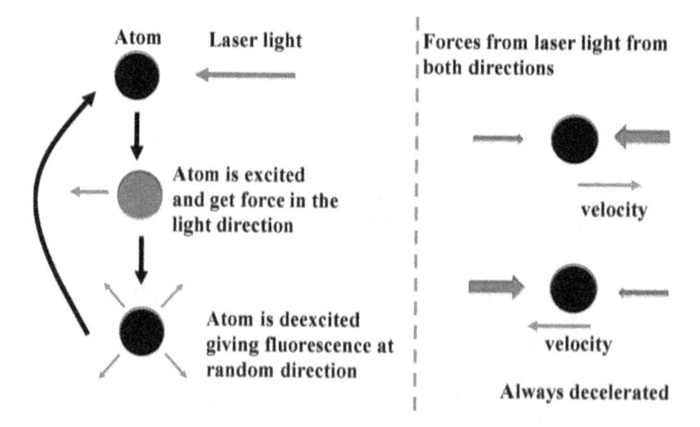

Figure 2.8. When an atom receives a force in the direction along the light direction with the cycle of laser induced excitation and deexcitation with spontaneous emission. Laser beams, with a frequency below the transition frequency, irradiate the atoms from opposing directions. Atoms interact only with laser light that opposes the motion of the atom and decelerate.

We assumed that the spontaneous emission transition from the excited state is possible only to the initial ground state. When spontaneous emission transition to another state is caused, the cooling cycle (absorption + spontaneous emission) stops afterwards. A repump laser, giving the transition to the excited state is needed to push atoms back to the cooling cycle, as shown in figure 2.9. With complex energy states, more repump lasers are required to maintain the laser cooling process. Laser cooling has been performed mainly for atoms with simple energy structures, such as alkali or alkali-earth atoms. Recently, laser cooling was successfully applied to several molecules with relatively simple energy structures [12, 13].

When laser cooling is performed in one direction, there is a heating effect in the other direction induced by the random force due to the spontaneous emission. To reduce the kinetic energy in two or three directions, four or six cooling lasers are required.

The magneto-optical trap (MOT) is an apparatus to trap laser cooled atoms in a small area [14]. Using an anti-Helmholz coil (a pair of two coils giving magnetic fields in counter propagating directions), a magnetic field proportional to the distance from the center is given. The transition frequency is shifted by the magnetic field (Zeeman shift) and the force from the laser light is proportional to the distance from the center as shown in figure 2.10. Therefore, the laser cooled atoms are trapped in a small area around the center.

Kinetic energy lower than the Doppler limit can be obtained using several methods, like polarization gradient cooling [15], sideband cooling [16], sideband Raman cooling [17] etc.

Evaporative cooling of trapped atoms is a useful method to reduce their kinetic energy below that attained by laser cooling [18]. With this method, atoms with energy higher than a certain energy (E_{max}) are removed. Energy distribution is out of the thermal equilibrium. Via elastic collision between atoms, the energy distribution

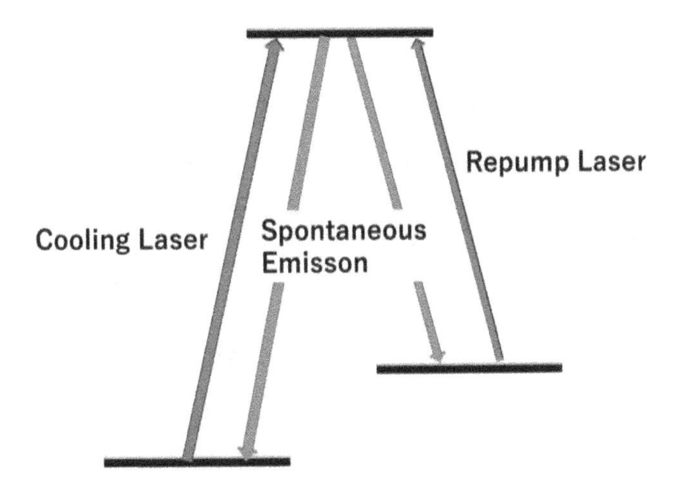

Figure 2.9. Maintaining the cooling cycle using a repump laser. Reproduced from [71] and [72]. Copyright IOP Publishing Ltd. All rights reserved.

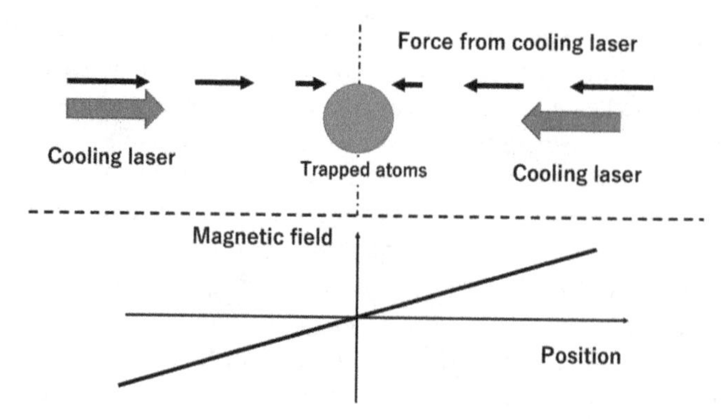

Figure 2.10. The fundamental concept of magneto-optical trap (MOT). Magnetic field is proportional to the distance from the center. Atoms get the trapping force from the cooling lasers, because of the Zeeman shift of the transition frequency. Reproduced from [73]. Copyright IOP Publishing Ltd. All rights reserved.

is transformed to a thermal equilibrium with a temperature lower than the initial one. With this procedure, some atoms get energy higher than $E_{\max}$ and are repelled. The energy distribution becomes equilibrium with a temperature (order of $E_{\max}/10k_B$), with which the distribution in the energy higher than $E_{\max}$ is negligibly small.

2.7 Ion trapping by RF electric field

This chapter introduces a method to trap ions in a small three-dimensional area using an inhomogeneous RF electric field. For an RF electric field $\vec{E}(\vec{r})\sin(\Omega t)$, trapping and expanding forces are periodically applied. Under certain conditions shown below, the time-averaged force can be a force to trap ions to confine in a three-dimensional small area (RF trap). The equation of motion of a charged particle is given by [19]

$$m\frac{\mathrm{d}^2\vec{r}}{\mathrm{d}t^2} = q_e\vec{E}(\vec{r})\sin(\Omega t). \tag{2.7.1}$$

When the change in position within the period of the RF electric field $\delta\vec{r}$ is negligibly small compared to $|\vec{r}|$ (change of $\vec{E}(\vec{r})$ is negligible small), the time-averaged force is obtained to be non-zero as follows (figure 2.11):

$$\vec{r} = \vec{r}_0 + \delta\vec{r}$$

$$m\frac{\mathrm{d}^2\delta\vec{r}}{\mathrm{d}t^2} = q_e\vec{E}(\vec{r}_0)\sin(\Omega t)$$

$$\delta\vec{r} = -\frac{q_e\vec{E}(\vec{r}_0)}{m\Omega^2}\sin(\Omega t) + c_1t + c_0$$

($c_{0,1}$: constant given by the initial position and velocity)

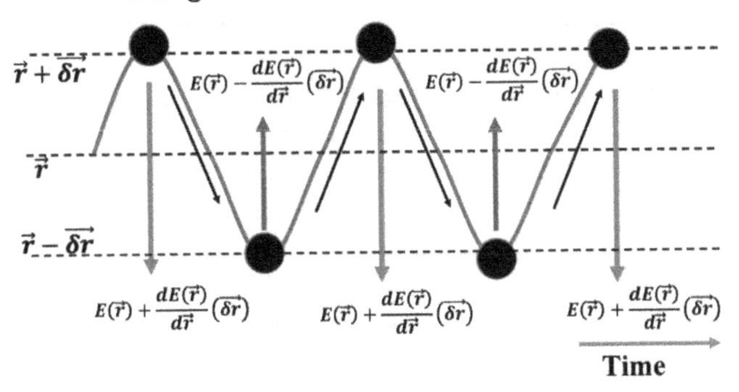

Figure 2.11. The mechanism to trap a charged particle using an AC electric field. The micromotion of the charged particle is synchronized to the AC electric field and a non-zero averaged force is induced in one direction. Reproduced from [19].

$$
\begin{aligned}
\left[m\frac{\mathrm{d}^2\vec{r_0}}{\mathrm{d}t^2}\right]_{ave} &= \left[q_e\vec{E}\left(\vec{r_0}+\vec{\delta r}\right)\sin(\Omega t)\right]_{\text{ave}} \\
&= \left[q_e\vec{E}\left(\vec{r_0}\right)\sin(\Omega t)\right]_{\text{ave}} + \left[q_e\frac{\mathrm{d}\vec{E}\left(\vec{r_0}\right)}{\mathrm{d}\vec{r}}\vec{\delta r}\sin(\Omega t)\right]_{\text{ave}} \\
&= -\left[q_e\frac{\mathrm{d}\vec{E}\left(\vec{r_0}\right)}{\mathrm{d}\vec{r}}\left\{\frac{q_e\vec{E}\left(\vec{r_0}\right)}{m\Omega^2}\sin(\Omega t)^2 + (c_1t + c_0)\sin(\Omega t)\right\}\right]_{\text{ave}} \\
&= -\frac{q_e^2\vec{E}\left(\vec{r_0}\right)}{2m\Omega^2}\frac{\mathrm{d}\vec{E}\left(\vec{r_0}\right)}{\mathrm{d}\vec{r}}.
\end{aligned}
\tag{2.7.2}
$$

The motion of ion (change of r_0) as represented by equation (2.7.2) is given as the motion with a pseudopotential field given by:

$$
P_{ps}(\vec{r_0}) = \frac{|q_e\vec{E}(\vec{r_0})|^2}{4m\Omega^2}. \tag{2.7.3}
$$

$P_{ps}(\vec{r_0}) \geqslant 0$ and there is a force to confine the ion at the position where $|\vec{E}(\vec{r_0})|$ is minimum. A large number of apparatus for ion trapping have been designed so that ions are confined at the position where $|\vec{E}| = 0$. Note that

$$
|\vec{E}(\vec{r_0})| \gg \left|\frac{\mathrm{d}\vec{E}(\vec{r_0})}{\mathrm{d}\vec{r}}\vec{\delta r}\right| \rightarrow \frac{q_e}{m\Omega^2}\left|\frac{\mathrm{d}\vec{E}(\vec{r_0})}{\mathrm{d}\vec{r}}\right| \ll 1 \tag{2.7.4}
$$

is required for the use of equation (2.7.2).

For ions trapped by the RF electric field, laser cooling can be performed with one cooling laser of frequency lower than the transition frequency. Trapped ions repeatedly undergo motion parallel and opposite to the direction of light and interact with the cooling laser only when the motion is in the opposite direction to the beam. The motion of the trapped ions in the three directions are coupled and cooled in all directions by a cooling laser in one direction. With this method, the kinetic energy of the trapped ion can be reduced to the Doppler limit $T_{\rm Doppler} = h\Gamma/4\pi k_B$ and the amplitude of the ion motion can be reduced to below 10^{-7} m (section 2.6). With the kinetic energy below 10^{-4} K, we must consider the discrete energy of trapped ions. Assuming that the pseudopotential is given by $P_{\rm ps} = m(2\pi\nu_t)^2 \left|\vec{r}\right|^2/2$,the motion energy of trapped ion is given by $E_{n_t} = (n_t + 1/2)h\nu_t$ (Appendix D), where n_t is integer. Using an electric dipole forbidden transition ($\Gamma/2\pi < \nu_t$), the $n_t \to n_t - 1$ transition can be induced. Using sideband Raman cooling [17] or EIT cooling [20], the ion motion energy can be reduced to the vibrational ground state $n_t = 0$.

Laser cooling of ions has been mainly performed with alkali-like ions (Ca^+, Sr^+, or Yb^+ ions) because of their simple energy structures. Other ions (molecular ions, highly charged ions etc) are co-trapped with laser cooled alkali-like ions and sympathetically cooled via the Coulomb interaction.

2.8 Atomic clocks using lasers and cold atoms

Laser cooling made it possible to reduce the kinetic energy of atoms. Using cold atoms, the interaction time between atoms and electromagnetic waves $\tau_{\rm int}$ becomes longer. The spectrum linewidth becomes narrower when it is given by $1/2\pi\tau_{\rm int}$. The frequency shift induced by the quadratic Doppler shift is also reduced.

The fountain-type Cs atomic clock was developed, and the fractional uncertainty of order 10^{-16} was obtained from several groups [21, 22]. Atoms that have been laser cooled and trapped are launched upward using laser light. The launched atoms pass through a microwave cavity, reverse direction due to gravity, and fall through the cavity again. The spectrum linewidth is given by the period between the first and second passage of the cavity, which is narrower than 1 Hz (larger than 100 Hz with atomic clocks using thermal beams). ^{87}Rb atomic clock with an atomic fountain was also developed with which a fractional uncertainty of order 10^{-16} was obtained [23].

Since the development of the frequency comb, the precision measurement of atomic transition frequencies in the optical region became a hot topic, aimed at a measurement uncertainty lower than that for the Cs atomic clocks. To observe the one-photon transition in the optical region without the first-order Doppler effect, the atoms are required to be localized in an area much smaller than the wavelength of the probe laser light (it is easy to localize within the microwave wavelength). Another requirement is that the force to confinement gives the same energy shifts at the upper and lower states of the transition, so that it does not induce the shift of the transition frequency. One method is the measurement of the transition of a single ion which is confined around the position where the RF electric field is zero. Reducing the amplitude of the vibrational motion by laser cooling, the shift in the transition frequency by the RF electric field (Stark shift) is negligibly small. The Coulomb force has no dependence on the quantum energy state, therefore, no shift in the transition frequency is induced. A fractional uncertainty lower than 10^{-17} was attained with the $^{27}Al^+$ ($S = 0, L = 0, J = 0$) → ($S = 1, L = 1, J = 0$) [24] and the $^{171}Yb^+$ ($S = 1/2, L = 0, J = 1/2$) → ($S = 1/2, L = 3, J = 7/2$) [25] transition frequencies. Here, S, L, and J are the quantum numbers of the electron spin, the electron orbital angular momentum, and the fine structure, respectively.

Another method was also developed to measure the transition frequencies of neutral atoms (figure 2.12). The atoms are laser cooled and trapped at the antinodes of the standing wave of the laser light (an optical lattice). Measurements are performed using many atoms to obtain a high signal-to-noise ratio; therefore, it is more advantageous to reduce the statistical uncertainty (Appendix C) rather than using a single trapped ion. The electric field of the trap laser light induces a shift in the transition frequency, which can be positive or negative depending on the frequency of the trap laser. Choosing the trap laser frequency (magic frequency), the shift in the transition frequency is zero as shown in figure 2.12. With the magic frequency, the trap potential energy at the upper and lower states are equal. The ($S = 0, L = 0, J = 0$) → ($S = 1, L = 1, J = 0$) transition frequency of ^{87}Sr, ^{171}Yb, and ^{199}Hg atoms atom have been measured with this method. The uncertainty of 10^{-18} has been obtained with the ^{87}Sr transition [26, 27]. The ratio of the ^{87}Sr and ^{171}Yb transition frequencies is given with a fractional uncertainty of 5×10^{-17} [28].

The attainable measurement uncertainty might be limited by the gravitational red shift, which gives the fluctuation by 10^{-18} with the change of altitude of 1 cm (sections 1.4 and 2.2). Therefore, we can estimate the difference of the altitude h_g from the comparison of atomic transition frequencies ν_a as

$$\Delta h_g = \frac{c^2}{g} \frac{\nu_a(h_g + \Delta h_g) - \nu_a(h_g)}{\nu_a(h_g)}, \tag{2.8.1}$$

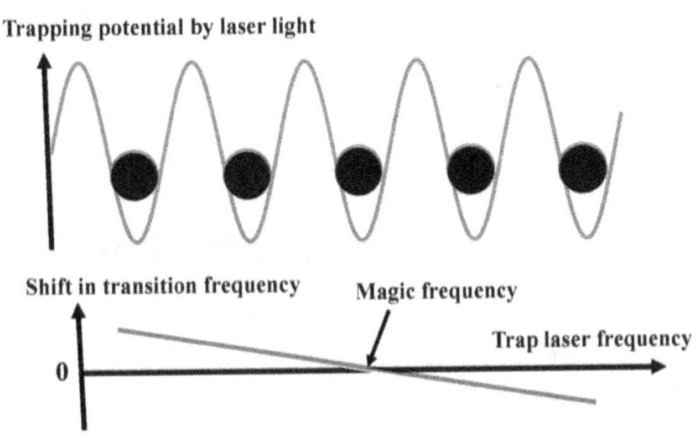

Figure 2.12. Concept behind the atomic lattice clock. Atoms are trapped at the antinodes of a standing wave of a laser light. The frequency of laser light is set where the shift in the transition frequency is zero (magic frequency).

where g is the gravitational acceleration and c is the speed of light. We can measure the difference of altitude at different positions with an uncertainty of several cm by measuring the atomic transition frequencies at each place with the uncertainties of 10^{-18} [29], which might offer new opportunities to explore seismology and volcanology [30].

Measurements of atomic optical transition frequencies at distant places are compared via laser light which is transformed via an optical fiber as shown in figure 2.13. The power loss of laser light in an optical fiber is low at the wavelength of 1.4–1.6 μm. The frequency stability of 7.7×10^{-17} was obtained with the fiber link of 2220 km distance [31]. To monitor the influence of the uplift of the Earth's surface, the comparison of atomic transition frequency should be performed in different places. The influence of the change of Earth's surface is not observed by just comparing the clocks at close places where the changes of the altitude are parallel. The temporal fluctuation induced by the Earth's tide (gravitational effect from the moon) is significant with the comparison between clocks at distant places (order of 10^{-16} with the comparison between different continents).

Here we introduce the development of a chip scale atomic clock (CSAC), which was possible after the development of a chip scale diode laser. We can measure the time and frequency with ultra-low measurement uncertainty. However, the apparatus of a conventional atomic clock is too large for use in our lives. The cost and the power consumption are also too high for personal use. Aiming for real use on our life, CSACs have been developed. Atomic clocks using optical transition frequencies cannot be chip scale, because several laser lights are required for probe, laser cooling etc. Therefore, only microwave transitions can be used to develop CSACs. However, the size of an apparatus using microwave irradiation cannot be smaller than its wavelength. To reduce the size, the microwave transition is observed using a laser with a chip scale. The microwave transition frequency between a_1 and a_2 states can be measured also using a laser with two frequency components. Laser light is absorbed by the atoms when its frequency corresponds to the transition frequency of either a_1–b or a_2–b (b is the highly excited state). However, the light is not absorbed when there are two frequency components, and their frequency interval equals the

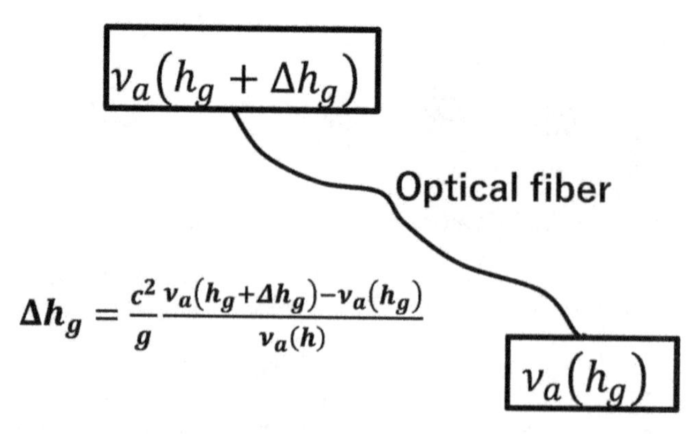

Figure 2.13. Measurement of the difference of altitude from the difference of the atomic transition frequency.

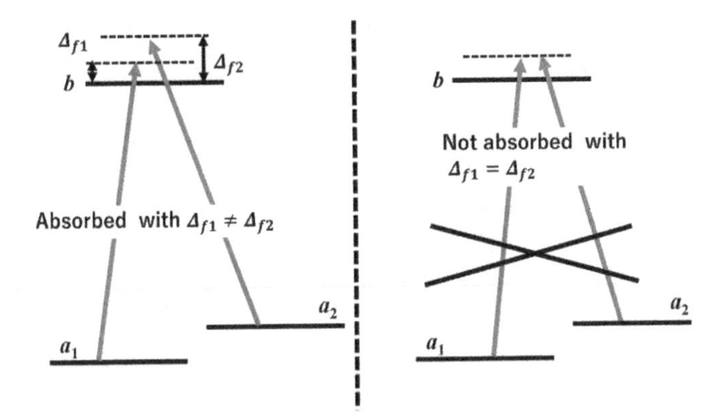

Figure 2.14. Fundamentals of electromagnetically induced transparency (EIT). Irradiating laser lights that are quasi-resonant to the a_1–b and a_2–b transitions with different detuning, both transitions are induced. However, both transitions are suppressed when both frequency detuning are equal.

$a_1 - a_2$transition frequency. This phenomenon is called electromagnetically induced transparency or EIT (figure 2.14), which is explained in detail in Appendix E.

The multiple frequency components given by $\nu(n_s) = \nu_c \pm n_s\nu_m$ (n_s is integer) are obtained modulating the laser frequency by $\nu = \nu_c(1 + \kappa_m \sin(2\pi\nu_m t))$. The microwave transition frequency is measured by monitoring ν_m at which the absorption of laser light is suppressed. With the development of the compact laser diode having low power consumption, the size of atomic clock using EIT can be much smaller than the microwave wavelength [32]. The uncertainty of measured frequency is induced by several components, and researchers are eager to suppress them. For example, spectrum broadening and shift of the transition frequency are induced by the interaction with laser light, which is significant when laser light is irradiated continuously. Pulsed lasers are used to suppress them [33]. Now the frequency stability of 10^{-12} with sizes smaller than 20 cm^3 have been attained.

2.9 Confirmation of quantum mechanics

2.9.1 Collision between cold atoms and molecules

The collision rate is given by

$$\Gamma_{\text{col}} = n_a v_{\text{rel}} \sigma_{\text{col}}, \tag{2.9.1}$$

where n_a is the atomic density, v_{rel} is the mean relative velocity, and σ_{col} is the collision cross-section. At first, we consider this with the semiclassical treatment. Assuming the classical straight-line track, the collision cross-section of is given by [34]

$$\sigma_{\text{col}} = 2\pi \int b P_{\text{col}}(b) db, \tag{2.9.2}$$

Where $P_{\text{col}}(b)$ denotes the probability of the collisional interaction with the closest distance (impact parameter) b. When the interatomic potential is ∞ at $r < d$, and 0 at $r > d$ (r: interatomic distance), $\sigma_{\text{col}} = \pi d^2$ is obtained from equation (2.9.2).

For the collision between cold atoms and molecules, the broadening of the atomic or molecular wave-packet is not negligible, and the collision cross-section is generally larger than that obtained using equation (2.9.2). Here, we analyze the elastic collision cross-section with the quantum treatment assuming a spherically symmetric potential. We consider the scattering of the incident plane wave, which is given by the coupling of partial waves with different angular momentum of the relative motion L (Appendix F), as follows [35]:

$$e^{ikz} = \sum i^L 2\sqrt{\pi(2L+1)}\frac{R_L(r)}{2\pi k}Y_L^0(\theta)$$

$$R_L(r) = (2\pi k)^{\frac{3}{2}} j_L(2\pi kr)$$

$$k = \frac{\sqrt{2\mu_r E}}{h}$$

E: collision energy, μ_r: reduced mass, j_L: spherical Bessel function

Y_L^0: see equation (1.5.39). (2.9.3)

The cross-section of each partial wave with a given L is given by

$$\sigma_L = \frac{2L+1}{\pi k^2} \tag{2.9.4}$$

and the collisional cross-section is given as:

$$\sigma_{\text{col}} = \sum \sigma_L P_{\text{col}-L} = \frac{1}{\pi k^2}\sum(2L+1)P_{\text{col}-L}, \tag{2.9.5}$$

where $P_{\text{col-}L}$ is the probability of collisional scattering with the angular momentum L. $P_{\text{col-}L}$ is obtained from the distortion of the wavefunction by the potential. Assuming a spherically symmetric potential, L does not change with the collision. When the potential works only at $r < a_p$, the region at $r > a_p$ is treated as the free space, and equation (2.9.3) is corrected as follows:

$$R_L(r) = \frac{(2\pi k)^{\frac{3}{2}}}{\sqrt{1+\beta_L^2}}\left[j_L(2\pi kr) + \beta_L n_L(2\pi kr)\right]$$

nL: Spherical Neumann Function. (2.9.6)

Equation (2.9.6) is applied only for $r > a_p$, therefore, it is not required to avoid the divergence at $r \to 0$ and the spherical Neumann function can be mixed. Comparing equations (2.9.3) and (2.9.6),

$$P_{\text{col}-L} = \frac{\beta_L^2}{1+\beta_L^2}. \tag{2.9.7}$$

For a large r,

$$j_L(2\pi kr) = \frac{\sin\left(2\pi kr - \frac{L\pi}{2}\right)}{2\pi kr}$$

$$n_L(2\pi kr) = \frac{\cos\left(2\pi kr - \frac{L\pi}{2}\right)}{2\pi kr} \tag{2.9.8}$$

and

$$\frac{1}{\sqrt{1+\beta_L^2}}\left[j_L(2\pi kr) + \beta_L n_L(2\pi kr)\right] = \frac{\sin\left(2\pi kr - \frac{L\pi}{2} + \delta_L\right)}{2\pi kr}$$

$$\beta_L = \tan(\delta_L)$$

$$P_{\text{col}-L} = [\sin(\delta_L)]^2$$

$$\sigma_{\text{col}} = \frac{1}{\pi k^2}\sum(2L+1)[\sin(\delta_L)]^2. \tag{2.9.9}$$

As an example, we consider a case of a potential well ($V(r) = 0$ and $r > a_p$ and $V(r) = \infty$ for $r < a_p$). Equation (2.9.9) is valid for $r > a_p$, and $R_L(r) = 0$ with $r < a_p$. Therefore,

$$j_L(2\pi ka_p) + \beta_L n_L(2\pi ka_p) = 0 \tag{2.9.10}$$

is required. Assuming $ka_p \ll 1$ (satisfied for atoms with kinetic energy lower than 1 mK),

$$\begin{aligned} j_L(2\pi ka_p) &= (2\pi ka_p)^L \\ n_L(2\pi ka_p) &= (2\pi ka_p)^{-(L+1)} \end{aligned} \tag{2.9.11}$$

and equation (2.9.10) is satisfied for

$$\beta_L = -(2\pi ka_p)^{2L+1} \tag{2.9.12}$$

and

$$\sigma_{\text{col}} = \frac{1}{\pi k^2}\sum(2L+1)\frac{(2\pi ka_p)^{4L+2}}{1+(2\pi ka_p)^{4L+2}}. \tag{2.9.13}$$

The contribution of the terms with $L \geqslant 1$, converge to zero with $k \to 0$ because of the centrifugal force potential. The collision cross-section with ultra-low kinetic energy is dominated by the term with $L = 0$ and it converges to $\sigma_{\text{col}} = 4\pi a_p^2$. This is four times larger than the term obtained from classical mechanics $\sigma_{\text{col}} = \pi a_p^2$.

When the interaction is weak, equation (2.9.9) is approximated as

$$\sigma_{\mathrm{col}} = \frac{1}{\pi k^2} \sum (2L+1) \delta_L^2$$

$$\delta_L = (\Delta\tau) \frac{2\pi}{h} (2\pi k)^3 \int \left[j_L(2\pi k r) \right]^2 V(r) r^2 dr, \tag{2.9.14}$$

where $\Delta\tau$ is the interaction time given by

$$\Delta\tau = \frac{\left(\frac{2\pi}{k}\right)}{\left(\frac{hk}{\mu_r}\right)} = \frac{2\pi\mu_r}{hk^2}. \tag{2.9.15}$$

This treatment is called the 'Born approximation'. For the interaction between electric dipole moments (potential is not spherical symmetric), $V(r) \propto r^{-3}$ and $\delta_L \propto k$. Therefore, the elastic collision cross-section is independent of the collisional kinetic energy, whereas the Born approximation is valid.

The quantum characteristic results from the interference between undistinguishable phenomena. For the collision between the identical particles in the same quantum state, there must be an interference of the relative positions $\vec{r}$ and $-\vec{r}$. For the collision between the same species, the wavefunction should be exchanged as follows:

$$\begin{aligned} \Psi_L(\vec{r}) &\to \frac{\Psi_L(\vec{r}) + \Psi_L(-\vec{r})}{\sqrt{2}} \quad \text{for Boson particles (with integer spin)} \\ \Psi_L(\vec{r}) &\to \frac{\Psi_L(\vec{r}) - \Psi_L(-\vec{r})}{\sqrt{2}} \quad \text{for Fermion particles (with half integer spin)} \end{aligned} \tag{2.9.16}$$

As $\Psi_L(-\vec{r}) = \Psi_L(\vec{r})$ for even L and $\Psi_L(-\vec{r}) = -\Psi_L(\vec{r})$ for odd L, equation (2.9.9) is replaced as

$$\begin{aligned} \sigma_{\mathrm{col}} &= \frac{2}{\pi k^2} \sum_{L=\mathrm{even}} (2L+1)[\sin(\delta_L)]^2 \text{ for Boson particles} \\ \sigma_{\mathrm{col}} &= \frac{2}{\pi k^2} \sum_{L=\mathrm{odd}} (2L+1)[\sin(\delta_L)]^2 \text{ for Fermion particles.} \end{aligned} \tag{2.9.17}$$

For the collision between the same species of Fermion particles, the term $L = 0$ does not exist and the collision effect induced by the short-range interaction is suppressed for the ultra-low kinetic energy. This effect has been experimentally observed for ^{83}K atoms [36]. This suppression is not observed when the elastic interaction is caused by the electric dipole–dipole interaction, because the collision term for $L \geqslant 1$ is also not suppressed for $k \to 0$ [37]. The inelastic collision (collisional transition to lower energy states) is suppressed also when it is caused by the electric dipole–dipole interaction [37].

2.9.2 Bose–Einstein condensation

With a thermal equilibrium state with the thermal dynamic temperature T, the distribution of energy E is proportional to $\Omega_s(E)\exp(-E/k_BT)$ as shown in Appendix A, which is derived assuming independent events (Ω_s: number of states). When atoms are cooled to ultra-low temperatures with high densities, atomic waves are broadened more than the interatomic distances, and the energy distribution cannot be treated as independent events. There is an interference between overlapped atomic waves as shown as (equations (1.5.54) and (2.9.16)).

$$\Psi(\vec{r}) \to \frac{\Psi(\vec{r}) + \Psi(-\vec{r})}{\sqrt{2}}$$

$$\Psi(0) \to \sqrt{2}\Psi(0) \quad \text{for Boson particles}$$

$$\Psi(\vec{r}) \to \frac{\Psi(\vec{r}) - \Psi(-\vec{r})}{\sqrt{2}}$$

$$\Psi(0) \to 0 \qquad \text{for Fermion particles.} \tag{2.9.18}$$

For Boson atoms (total spin is integer), the atomic waves have positive interference and the square of the amplitude of wavefunction of two atoms is double at $\vec{r} = 0$. Boson atoms tend to have uniform quantum states and condense to the lowest energy, as shown in figure 2.15. This phenomenon is called Bose–Einstein condensation (BEC). The phases of atomic waves in BEC are uniform, and the group of atoms are treated like a single mechanical entity with a wavefunction on a macroscopic state. The first atomic BEC was made with Rb atoms in a magnetic trap with the kinetic energy 0.17 μK (laser cooling + evaporative cooling) in 1995 [38]. BEC was attained with Na atoms in an optical trap in 1998 [39].

The following phenomena were observed with atoms in the BEC state.

(1) The slow light propagation (17 m s^{-1}) was observed in BEC, because of a very large dielectric constant [40].

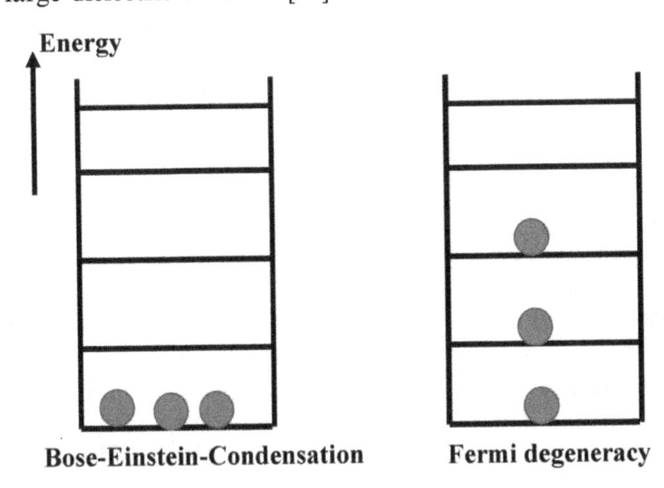

Figure 2.15. Energy distribution with Bose–Einstein condensation and Fermi degeneracy.

(2) The BEC state is stable (no phase transition to the liquid or solid) when the interatomic potential is repulsive. When it is switched to attractive (by switching the magnetic field etc), the size of the BEC rapidly becomes smaller. The collapse of BEC stops and expansion starts when the expansion by the momentum uncertainty overcomes the attractive force. After the expansion of BEC, a small BEC remains, in just the way a neutron star is made after the explosion of a star [41].

(3) The coherent atomic beam called the atomic laser was attained, just like laser light as the BEC state of a photon. A coherent Na atomic beam looking much like a dripping tap was obtained by letting gravity pull off partial pieces of atoms in the BEC state when the trapping force was periodically released [42].

Particles with the spin quantum number of half integer are Fermion, with which atomic waves have negative interference (wavefunction is zero at $\vec{r} = 0$) as shown in equation (2.9.18). Therefore, only one particle can exist in a quantum state (Pauli exclusion principle). With the group of Fermion particles, energy distribution cannot be localized in the lowest state also with ultra-low temperature, but they can occupy the states in order from the lowest, as shown in figure 2.15. The simplest example is electrons in atoms. From the energy gap between principal quantum number $n = 1$ and 2 state, all electrons seem to be localized in the $n = 1$ state from the statistical mechanical consideration (Appendix A). However, only two electrons are in the $n = 1$ state, which is satisfied only with H and He atoms. With the Li atom having three electrons, two electrons are in the $n = 1$ state and one electron is in the $n = 2$ state.

However, Fermion atoms can pair up with opposite spins and have a Boson-like states (Cooper pair). Paired Fermion atoms can also undergo the BEC state. The ^{3}He and ^{4}He atoms at low temperature flow with zero friction (superfluidity). This is because the atoms form a BEC state, with which the motion of all particles are macroscopic and any scatterings of a small fractions of particles are also forbidden if there are impurities. Superconductivity is the superfluidity of electrons forming Cooper pairs.

It is useful to observe the phenomena of atoms in the BEC state trapped by a standing wave of a laser light, because it gives an analogy of periodic potential for the electrons in a solid crystal. The laser light gives a strict periodic potential while there are lattice defects in a solid crystal. It is also another merit that we can control the parameters freely. With low potential depth, atoms are free to move to the neighboring lattice, as shown in figure 2.16. This is the superfluidity with atoms in the BEC state, which gives an interpretation of the superconductivity of a solid crystal. With a potential depth higher than a certain value, atoms cannot move to another lattice because of the repulsive interatomic interaction (figure 2.16). With this mechanism, some conductive materials can be transformed to insulators with low temperature (Mott insulator). There is a research trapping ^{174}Yb (Boson) and ^{6}Li atoms (Fermion) simultaneously in an optical lattice. The ^{6}Li atoms in an optical lattice have the role of electrons in a crystal, and ^{174}Yb atoms have the role of the

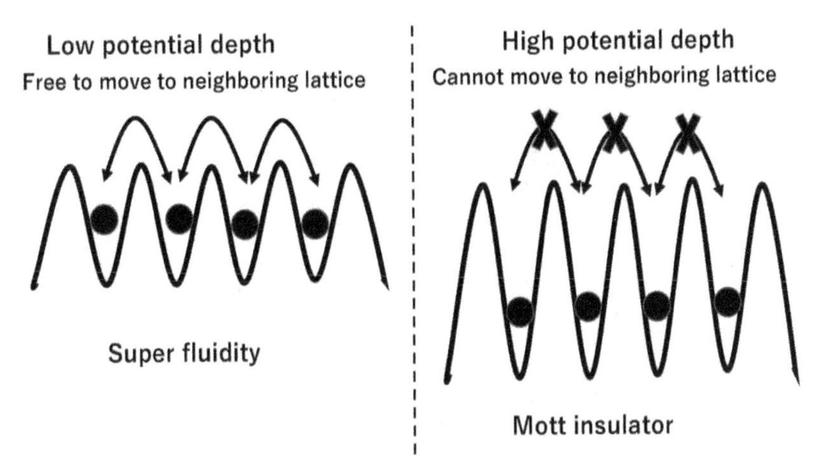

Figure 2.16. Phase transition between the superfluidity and the Mott insulation. Reproduced from [73].

impurity [43]. There is a phenomenon called 'Band insulator' with Fermion atoms; atoms cannot move to the neighboring lattice occupied by another atom. This phenomenon looks like a Mott insulator, but is caused by the Pauli exclusion principle prohibiting multi-Fermion atoms in the same state [44].

There is a project to attain the BEC with positronium (Ps) [45]. Ps is a bound state of an electron and a positron, and it is the lightest atom. Its energy structure is much simpler than atoms because it is given only by the electromagnetic force. The lifetime of Ps in the $(n, L) = (1,0)$ state (n: principal quantum number, L: rotational quantum number) is short because there is an area where the wavefunctions of electron and positron overlap and pair annihilation is caused. The total spin of electron and positron is 1 or 0. In the spin state of 1, there is a magnetic repulsion force between electron and positron and the lifetime (1.42×10^{-7}s) is much longer than that in the spin state of 0 having the magnetic attractive force (1.25×10^{-10}s). The BEC state might only be attained with the spin of 1. With the BEC state, the decay process is uniform for all Ps. The rate of simultaneous decay of all Ps in the BEC state is much lower than that for a single Ps. Observing the annihilation by the interaction between electron and positron at the overlapping position, we can get information about the interaction between the nucleus and electron in atoms. Electrons in the $L = 0$ state have non-zero distribution at the position of nuclear and they can interact with the nucleus by the weak nucleus force, which violates the parity symmetry in the atomic molecular energy structure (section 3.14). Irradiating a microwave of 203 GHz, the transition of the spin state $1 \rightarrow 0$ is induced and it becomes a source of γ ray with a uniform phase.

The BEC state of Ps is attained with 14 K, when the density is 10^{17} cm^{-3}. To attain this condition within the Ps lifetime, Ps should be created with densities higher than 10^{17} cm^{-3} within 5×10^{-8} s and laser cooled within 3×10^{-7} s. The University of Tokyo project [45] produced Ps by the collision of positrons on the surface of the converter (SiO_2 aerogel), cooled to 4 K by a cryogenic refrigerator (figure 2.17). The

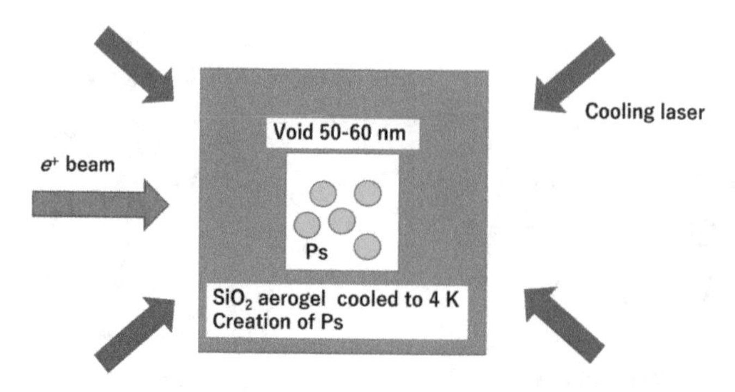

Figure 2.17. The system of Ps creation, pre-thermalization by the converter (SiO_2 aerogel) and laser cooling in the void. Reproduced from [72]. Copyright IOP Publishing Ltd. All rights reserved.

thermalization of Ps to 100 K is already attained. Ps laser cooling is performed using s laser light with a wavelength of 243 nm. The temperature of 10 K is expected to be attained within 5×10^{-8} s [46]. The pulsed cooling laser (length 300 ns) should have a spectrum broadening of 460 GHz, so that the cooling is effective to Ps with a wide velocity area. The University of Tokyo group prepared a cooling laser satisfying these requirements. The Ps BEC is expected to be attained in the near future.

2.9.3 Entangled state between two particles

One of the most special quantum phenomena is the existence of the 'superposition' of multiple states. Now we discuss the superposition of the two eigenstates of two particles A and B

$$\begin{aligned}\Phi_A &= c_1\phi_1 + c_2\phi_2\\ \Phi_B &= c_\alpha\phi_\alpha + c_\beta\ \ \phi_\beta.\end{aligned} \tag{2.9.19}$$

By the state measurement, the superposition is transformed to one of the eigenstates with the possibility of $|c_x|^2$ ($x = 1, 2, \alpha, \beta$). With the independent events, the total state is given by

$$\Phi_{\text{tot}} = \Phi_A\Phi_B = c_1c_\alpha\phi_1\phi_\alpha + c_1c_\beta\phi_1\phi_\beta + c_2c_\alpha\phi_2\phi_\alpha + c_2c_\beta\phi_2\phi_\beta \tag{2.9.20}$$

However, there is also a state called the 'entangled state'

$$\Phi_{\text{ent}} = C_{1\alpha}\phi_1\phi_\alpha + C_{2\beta}\phi_2\phi_\beta. \tag{2.9.21}$$

Measuring the state of A from the entangled state, we know the state of B simultaneously. Quantum teleportation is a technology which allows us to know information about particle B just by the measurement of particle A [47]. With this method, signal transduction faster than the speed of light is possible.

The entangled states have been produced often with two co-trapped ions. The vibrational motions of both ions in the trapping potential are not independent; when the motion of one ion is excited, the motion of the other ion is also excited by the

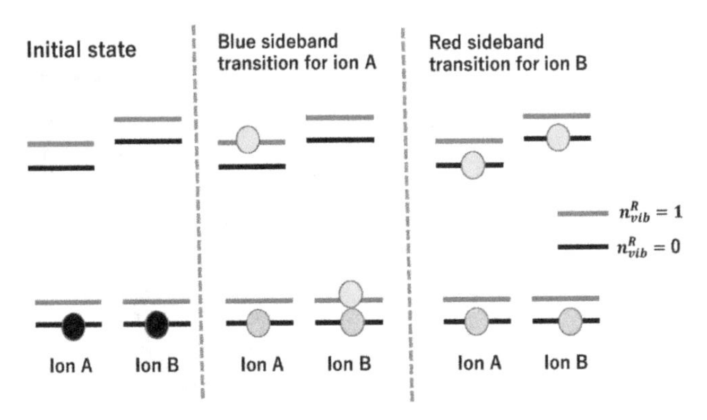

Figure 2.18. One example to produce an entangled state using a blue sideband transition of ion A and red sideband transition of ion B. Reproduced from [19].

Coulomb interaction. As shown in Appendix D, the eigenvalues of the motion energy are given by $E^R_{m-\mathrm{vib}} = (n^R_{\mathrm{vib}} + 1/2)h\nu^R_m$ (n^R_m: integer, ν^R_m: frequency of the vibrational motion. The entangled state is given with the following procedure for example (see figure 2.18).

(1) As the initial state, $\Phi = \langle \Phi_A | n^R_{\mathrm{vib}} | \Phi_B \rangle = \langle \phi_1 | 0 | \phi_\alpha \rangle$ is prepared.

(2) With ion A, the $1 \rightarrow 2$ blue sideband transition (motion mode is simultaneously excited) is induced by a laser light with a frequency higher than the $1 \rightarrow 2$ transition frequency by ν^R_m: $\Phi = c_1\langle \phi_1 | 0 | \phi_\alpha \rangle + c_2\langle \phi_2 | 1 | \phi_\alpha \rangle$

(3) A laser light with a frequency lower than the $\alpha \rightarrow \beta$ transition frequency by ν^R_m induces the $\langle \phi_2 | 1 | \phi_\alpha \rangle \rightarrow \left\langle \phi_2 | 0 | \phi_\beta \right\rangle$ transition, but no transition is induced from $n^R_{\mathrm{vib}} = 0$: $\Phi = c_1\langle \phi_1 | 0 | \phi_\alpha \rangle + c_2\left\langle \phi_2 | 0 | \phi_\beta \right\rangle$

The entangled state is used, e.g., to monitor the quantum state of ion, with which the fluorescence from a cycle transition (excitation + spontaneous emission transition) is not observed. Making an entangled state with a co-trapped alkali-like ion, the state of the target ion can be monitored by observing the fluorescence from the alkali-like ion.

2.9.4 'Schrödinger's cat' state

A quantum state can be a coupled state between states with different physical values. The 'Schrödinger's cat' state is a coupled state between multiple physical values with the difference not negligible also with macroscopic sight.

Monroe *et al* observed the Schrödinger's cat phenomenon using a single ion trapped by an RF electric field (section 2.7) as shown in figure 2.19 [48]. The ion is prepared at the state of coupling with two different quantum energy states 'a' and 'b'. Using two laser lights with counter propagating directions and a slight frequency difference, a pushing force to a positive (negative) direction is given to the ion in the 'a (b)' state. The vibrational motion in the 'a' state is $x = x_0 \sin[2\pi\nu_v t]$ (called the Vib_1 state) and the 'b' state $x = -x_0 \sin[2\pi\nu_v t]$ (called the Vib_2 state) [48].

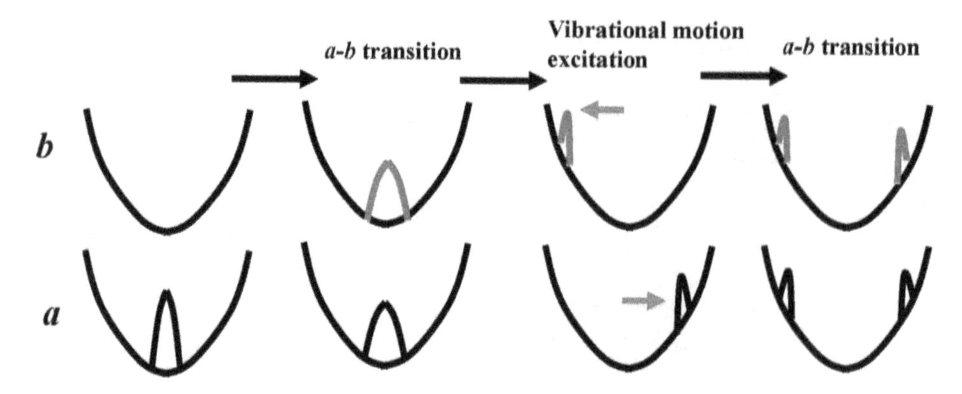

Figure 2.19. The procedure to observe the 'Schrödinger's cat' phenomenon. The state is initially localized to the state 'a'. The coupled state between the 'a' and 'b' states is constructed by the a–b transition. Then the vibrational motion is excited using two laser light, which gives forces to the 'a' and 'b' states in the opposite directions. Giving the a–b transition again, both vibrational motion modes are coupled in the 'a' and 'b' states. Reproduced from [19].

The amplitude of the vibrational motion x_0 is much larger than the broadening of the wave-packet, then the wave-packet is localized at two distant places (maximum $2x_0$) simultaneously. With this state, the vibrational mode (Vib_1 or Vib_2) is determined by measuring the energy state, because the energy state and the vibrational mode are entangled. Giving a – b transition again, both states are mixed. After that, both vibrational modes are coupled also after the measurement of the energy state. When both parts of the wave-packet are overlapped at $x \approx 0$, an interference is observed, showing that we cannot distinguish which part of the wave-packet is real. The 'Schrödinger's cat' state exists, but it is difficult to observe in real life because it decays within a period shorter than 10^{-5} s [48].

The Schrödinger's cat state is possible, not only with the one-dimensional vibrational motion, but also with the rotational motion in the counter propagating directions. The overlapping of two parts of wave-packet is observed as the interference [49].

2.9.5 Quantum chaos

The interpretation of quantum mechanics is mostly established, but there is still a paradox which is not solved yet. In principle, all phenomena which is described by Newtonian mechanics are also described by quantum mechanics, although the wavelength of material waves are negligibly small for the object much larger than atomic size. But there are phenomena which cannot be described with the current interpretation of quantum mechanics.

There is a phenomenon called 'chaos', which means 'unpredictable phenomenon in future'. There are equations to predict future phenomena but the prediction of the future is not possible from the solution when it changes drastically by a slight difference of the initial condition. For example, there are equations to predict the weather, but its solution changes drastically with slight differences of circumstance.

The chaos is derived from the non-linearity of the differential equation. However, these temporal expansions are not derived from the Schrödinger or Dirac equations, which are linear [50].

Another discrepancy is that the fundamental of chaos is the sensitivity to the slight difference of the initial condition. In quantum mechanics, no difference of phenomena should be derived with the difference of any physical values within the uncertainty principle.

Chaotic motion exists in reality. Several approaches have been employed to describe the quantum motions:

(1) development of methods to solve Schrödinger equations with an intensive perturbation;
(2) correlating statistical descriptions of energy eigenvalues with the classical behavior of the same Hamiltonian;
(3) study of probability distribution of eigenstates;
(4) use of the semiclassical method to connect the classical trajectory of the dynamic system with quantum features.

No final answer has been obtained.

2.9.6 Real use of the quantum effect

(1) Inertial navigation system

In the the Age of Discovery (15th to 18th centuries), navigators found it necessary to measure their positions. Latitude was measured from the height of Polaris above the horizon. Longitude was measured from the position of constellations at fixed times and development of accurate clocks for the use on a ship was required for this purpose. Now we can measure our position (latitude, longitude, altitude) using the Global Positioning System (GPS, developed by US military in 1973) [51]. The fundamental of GPS is the measurement of the distance between the receiver and the satellites from the propagation time of the radio wave. In principle, we can determine the position by measuring the distances from three satellites. For precision measurement of the propagation time, atomic clocks are used in satellites. The clock in the receiver is currently a crystal clock because the size of atomic clocks cannot be so compact (section 2.8 indicates CSAC). The error by the clock in the receiver is corrected by measuring the distances from four satellites. We can always measure position using more than seven satellites.

However, the radio waves from satellites cannot be received in a submarine cruising at an abyssal. In this case, an inertial navigation system is used [52]. This system measures the acceleration (using an accelerometer) and rotation angular velocity (using gyroscopes). The velocity is given by the time integral of acceleration by time, and the position is given by the time integral of the velocity. The motion direction against the earth's magnetic field is also monitored using magnetometers.

The interference signal between waves in different paths is used for the gyroscope and accelerator. The interference signal between both waves indicates the phase difference between two paths. Focusing after the decimal point of the phase

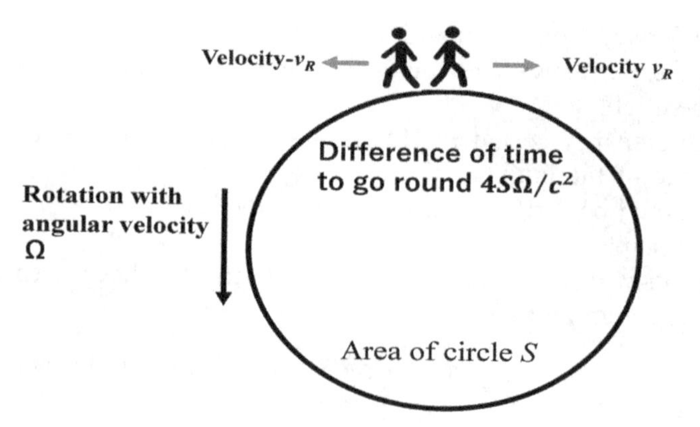

Figure 2.20. The Sagnac effect, which gives the time difference of the time to go around the rotating circle because of the relativistic effect.

difference divided by 2π, a slight change of the propagation time or frequency can be detected with high sensitivity.

To measure the rotation angular velocity Ω with the gyroscope the Sagnac effect is used. The Sagnac effect is a relativistic effect, which makes the time to go around a rotating circle (radius R_0, area S) depend on the path as shown in figure 2.20. With the motion with a velocity v_R parallel and antiparallel to the rotation, the time to go around the circle is given by

$$\text{parallel velocity } v_R + R_0\Omega \text{ time to go around the circle } T_+ = \frac{2\pi R_0}{v_R\sqrt{1-\left(\frac{v_R+R_0\Omega}{c}\right)^2}}$$
$$\text{antiparallel velocity } v_R - R_0\Omega \text{ time to go around the circle } T_- = \frac{2\pi R_0}{v_R\sqrt{1-\left(\frac{v_R-R_0\Omega}{c}\right)^2}}, \quad (2.9.22)$$

and the difference of the time is given by

$$\delta T = T_+ - T_- = -\frac{4\pi R_0^2\Omega}{c^2} = -\frac{4S\Omega}{c^2}. \quad (2.9.23)$$

This time difference can be measured, e.g., from the interference between ring type laser lights propagating in both directions. Observing the interference of the lights (frequency ν) propagating both directions, the phase shift of

$$\phi_{Gy} = 2\pi\nu(\delta T) = -\frac{8\pi\nu S\Omega}{c^2} \quad (2.9.24)$$

is observed. Using a gyroscope with a ring laser, the attainable accuracy is of the order of 10^{-3}.

Equation (2.9.24) shows that the Sagnac effect is detected with higher sensitivity with higher frequency. Atomic interferometry is useful to detect the slight difference of the propagation time because the frequency of atomic wavefunction is given by

mc^2/h, which is much higher than the frequency of laser light. The atomic interferometry is applied also for the gyroscope, with which the uncertainty of the rotational angular velocity is reduced to 10^{-12} [53]. Figure 2.21 indicates the system of a gyroscope using a beam of atoms having states 'a' and 'b' (the initial state is a).

The first laser light in the direction perpendicular to the atomic beam induces the a → b transition by 50% at the time $t = 0$ ($\pi/2$ transition shown in Appendix E). When the transition is caused, the atom also gets the momentum of the laser light and velocity component in the direction parallel to the laser light changes. After a certain period $t = \tau_0/2$, atoms interact with the second laser light to induce the a ↔ b is caused by 100% (π transition). The third interaction with laser light at $t = \tau_0$ makes the mixture of atoms from two paths ($\pi/2$ transition). The phase difference observed by the interferometry is given by $\phi_+ = 2\pi\left[(\nu_s + \delta\nu_s)(\tau_0 + \tau_{\rm sag}) - \nu_s\tau_0\right] \approx 2\pi\left(\nu_s\tau_{\rm sag} + \tau_0\delta\nu_s\right)$, where $\tau_{\rm sag}$ the difference between the time to arrive at the third interaction area given by the Sagnac effect and $\delta\nu_s$ is the variation of the frequencies of atomic wavefunction ν_s induced by the inertial force of the acceleration. Giving the inverse direction of the atomic beam, the phase shift is $\phi_- = 2\pi\left(\nu_s\tau_{\rm sag} - \tau_0\delta\nu_s\right)$. The rotational angular velocity and the acceleration are obtained from $\left(\phi_+ - \phi_-\right)/2$ and $\left(\phi_+ + \phi_-\right)/2$, respectively. Taking a long interaction time τ_0 by using laser cooled atoms, the acceleration can be measured with high accuracy.

A gyroscope can be constructed also using a single trapped ion in a Schrödinger cat state (one ion can be in two motion modes simultaneously, as shown in section 2.9.4). When the trapped ion can have a circular motion in two opposite directions, and we can observe interferometry when two parts are overlap, which indicates the influence of the Sagnac effect [49]. A compact gyroscope can be constructed using a trapped ion.

(2) Quantum computer

The fundamental of the computer is the 'bit', giving two positions '0' or '1'. With the classical computer, these positions are definite. When there are inputs of n_b bits,

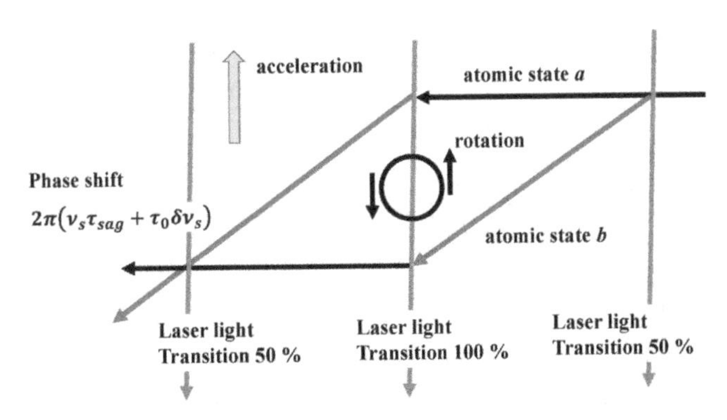

Figure 2.21. The interferometry of the atomic beam to detect the rotational angular velocity and the acceleration. Using laser system, the atomic path is split into two parts and the phase difference between two paths is given by the difference of the averaged frequency $\delta\nu_s$ (induced by acceleration force) and the time to arrive at the position to cause the interference (induced by the Sagnac effect).

calculation of 2^{n_b} times (number of combinations) is required for the classical computer. With a quantum computer, the bit is called a 'qbit' which can have a superposition (simultaneous existence in multiple states) of '0' and '1' states and all combinations can be calculated with one procedure as shown in figure 2.22. The calculation time and the power consumption are drastically reduced in comparison with classical computers

The idea of the quantum computer was given by Benioff [54] and Shor [55]. In 1992, Deutsch and Jozsa developed the first algorithm with which a quantum computer can give a rapid solution [56]. In 1994, Shor developed 'Shor's algorithm', with which prime factorization (not realistic with a classical computer) could be performed in a very short time [57]. Since the development of 'Shor's algorithm', the quantum computer became a hot topic. An algorithm was also developed for the error correction [58]. 'Grover's algorithm' was also developed for universal use [59]. The main problem is that the superposition at micro size (atoms or molecules) is destroyed with a slight fluctuation of the circumstance (decoherence). Decoherence was proved experimentally [60], and it creates a barrier for the development of the quantum computer. The entangled state of trapped ions was expected to be advantageous for the quantum computer because they are isolated from the circumstance [61].

All computational operations are combination of the controlled-NOT (CNOT) gates with two *q*-bits (control bit and target bit) [62]. With a CNOT gate, the following operation is performed with the target bit depending on the control bit.

Control bit 0 Target bit $0 \to 0, 1 \to 1$

Control bit 1 Target bit $0 \to 1, 1 \to 0$

This gate was realized by Monroe *et al* using a $^9Be^+$ ion, using the motion energy mode n_{vib} (= 0 or 1, Appendix D) as the control bit and the hyperfine energy state of the ion $\downarrow$ $(= |F = 2, m_F = 2\rangle)$ and $\uparrow$ $(= |F = 1, m_F = 1\rangle)$ as the target bit. An additional state aux $(= |F = 2, m_F = 0\rangle)$ is also used to give a phase transition [63].

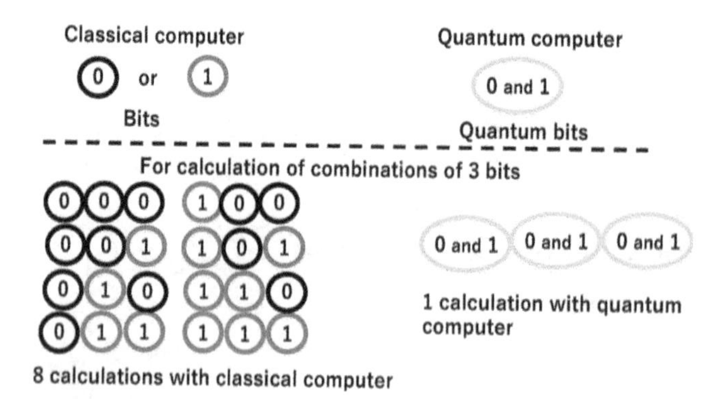

Figure 2.22. The 3 bits of classical and quantum computers. Eight times calculations are required with a classical computer to cover all possible combinations of 0 and 1 with each bit. With a quantum computer, the bit (qbit) can have a coupling of '0' and '1' states, and we can get the result with one calculation. Reproduced from [72] and [19].

The CNOT gate is given using the following procedure. The phase procedure of the transition is shown in Appendix E.

(1) $\pi/2$-carrier transition

(2) 2π-blue sideband transition $\langle n_{\text{vib}}, \uparrow\rangle \rightarrow \langle n_{\text{vib}} - 1, \text{aux}\rangle$ to make the $\langle 1, \uparrow\rangle \rightarrow -\langle 1, \uparrow\rangle$ transition: the transition phase after this procedure $\pi/4$ with $n_{\text{vib}} = 0$ and $5\pi/4$ with $n_{vib} = 1$

(3) $-\pi/2$-carrier transition: the phase transition $\pi/4 \rightarrow 0$ for $n_{\text{vib}} = 0$ and $5\pi/4 \rightarrow \pi$ for $n_{\text{vib}} = 1$.

With this procedure,

$$\langle 0, \downarrow\rangle \rightarrow \frac{\langle 0, \downarrow\rangle + \langle 0, \uparrow\rangle}{\sqrt{2}} \rightarrow \frac{\langle 0, \downarrow\rangle + \langle 0, \uparrow\rangle}{\sqrt{2}} \rightarrow \langle 0, \downarrow\rangle$$

$$\langle 0, \uparrow\rangle \rightarrow \frac{\langle 0, \downarrow\rangle + \langle 0, \uparrow\rangle}{\sqrt{2}} \rightarrow \frac{\langle 0, \downarrow\rangle + \langle 0, \uparrow\rangle}{\sqrt{2}} \rightarrow \langle 0, \uparrow\rangle$$

$$\langle 1, \downarrow\rangle \rightarrow \frac{\langle 1, \downarrow\rangle + \langle 1, \uparrow\rangle}{\sqrt{2}} \rightarrow \frac{\langle 1, \downarrow\rangle - \langle 1, \uparrow\rangle}{\sqrt{2}} \rightarrow \langle 1, \uparrow\rangle$$

$$\langle 1, \uparrow\rangle \rightarrow \frac{\langle 1, \downarrow\rangle + \langle 1, \uparrow\rangle}{\sqrt{2}} \rightarrow \frac{\langle 1, \downarrow\rangle - \langle 1, \uparrow\rangle}{\sqrt{2}} \rightarrow \langle 1, \downarrow\rangle$$

The action of this gate entangles the ions, making it a fundamental operation for the construction of an arbitrary quantum computation among many ions. The gate does not work perfectly because of laser intensity fluctuation or noisy ambient electric field. Although it is not realistic to make an entangled state with many ions (more than 20) from the motion energy mode, the entanglement is constructed using photons emitted from ions [64]. The photons can travel through an optical fiber and can make an entangled state between ions at distant places.

(3) Atomic magnetometer

Atomic magnetometers work by measuring the precession frequency of certain atoms having a magnetic dipole moment in a magnetic field. Alkali atoms are useful for this purpose, because of an unpaired electron with the spin of 1/2. The precession frequency is proportional to the magnetic field being measured. Since the frequency can be measured with a high accuracy, we can determine the magnetic field with high accuracy.

Irradiating a right (left) circular polarized laser light in the x-direction to alkali atoms having an electron spin quantum number $S = 1/2$, all atoms are pumped to the $S_x = 1/2$ $(-1/2)$ state by the torque given by the rotation of the light magnetic field. When there is a magnetic field, for example in the z-direction, the information about S_x is no longer valid and the precession of alkali atoms is induced. When a linear polarized laser is irradiated in the y-direction, the polarization direction changes by the precession of electron spin [65] as shown in figure 2.23. The sensitivity of 0.16 fT $\text{Hz}^{-1/2}$ was obtained using K atoms [66]. The transition between the different

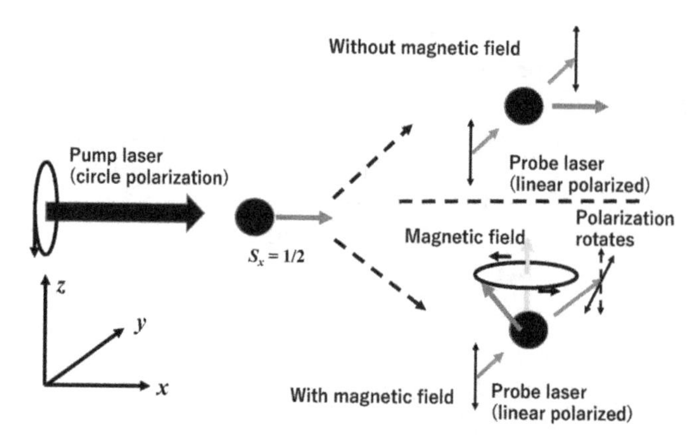

Figure 2.23. Fundamentals of atomic magnetometer. The spin in the x-direction is localized using a circularly polarized laser. When there is a magnetic field in the z-direction, precession of atoms is induced. When a linear polarized laser light is irradiated in the y-direction, the polarization direction changes by the precession. Reproduced from [73].

spin state can be induced also by the interatomic collision, which makes the noise on the measurement signal.

Using alkali atoms in the spin polarized BEC state, the temporal and special resolution is improved drastically [67]. Using alkali atoms in the BEC state, the phase of the procession is uniform for all atoms. Illuminating a circularly polarized laser light to atoms in the BEC state, the image of the procession is directly observed using a camera. Using ^{87}Rb atoms in the BEC state, the sensitivity of 0.15 pT Hz$^{-1/2}$ was obtained for unity duty cycle measurement. The magnetometer using atoms in the BEC state made it possible to detect the magnetic field with the uncertainty 10^{-7} times lower than the earth's magnetic field. The high resolution of the position is useful to get a human magneto cardiograph (MGC).

2.10 Detection of gravitational wave

With the theory of general relativity, gravity was described as a distortion of space and propagates as a wave with the speed of light (section 1.4). For gravitational waves, the size in the direction perpendicular to the propagation repeatedly expands and contracts, as shown in equation (1.4.2). This effect has been very difficult to observe, because the change in size is so small (ratio of 10^{-21}). However, a phenomenon was discovered to confirm the existence of gravitational waves indirectly. The orbital period of a binary neutron star PSR 1913 + 16 (7.75 h) decreased by 7.65×10^{-5} s/year [4]. They are expected to collapse 300 million years from now. This result shows the loss of energy in orbital motion, and it is reasonable to state that this energy loss is in the form of radiating gravitational waves. The measured variation in the orbital period is in good agreement with the estimation obtained from the theory of general relativity to within 0.2%.

A laser interferometer was constructed for the direct detection of gravitational waves. The apparatus (figure 2.24) consists of a laser (Nd: YAG laser, wavelength:

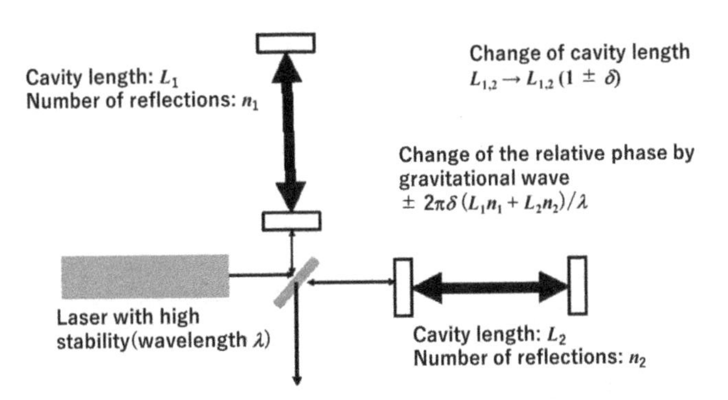

Figure 2.24. Laser interferometry system to detect gravitational waves. Reproduced from [71–73]. Copyright IOP Publishing Ltd. All rights reserved.

$\lambda = 1064$ nm) that shines light onto a half mirror, which splits and redirects the light into two perpendicular directions. In the arms of the interferometer (resonators of length $L_{1,2}$), the light is reflected many times ($n_{1,2}$ times). Both laser beams are brought together in the half mirror again to form an interference pattern exhibiting a phase difference of

$$\phi = 2\pi(n_1L_1 - n_2L_2)/\lambda. \tag{2.10.1}$$

When the gravitational wave changes the cavity length by $L_{1,2} \rightarrow L_{1,2}(1 \pm \delta)$, there is a change in phase difference of

$$\delta\phi = 2\pi\delta(n_1L_1 + n_2L_2)/\lambda. \tag{2.10.2}$$

The frequency stability (10^{-15}) of the laser light was important in attaining high sensitivity to the slight changes in cavity length. The laser power also must be stabilized to 10^{-6}, because the fluctuation in radiation pressure changes the cavity length.

The first gravitational wave was observed on 14 September 2015, at the Laser Interferometer Gravitational-wave Observatory (LIGO) in Hanford (Washington, US) and Livingston (Louisiana, US) [68]. Each has a sensitivity to detect the change in the length of the 4 km cavity of 1000th of the width of a proton (10^{-18} m). There is a slight difference with the detection times of the gravitational wave at both laboratories, with which we can find the position of the source of the gravitational wave. The gravitational wave was caused by the merger of a pair of black holes of 36 and 29 solar masses at a place 1.3 billion light years distant. The frequency was at first 35 Hz and it was raised to 210 Hz. The motion of black holes became faster as the get closer to each other.

The system in LIGO was reconstructed to improve the detection sensitivity. Gravitational waves have been since been observed often afterwards. Now the gravitational wave has also been observed in VIRGO (Italy) [69]. KAGRA (Japan) has also improved detection sensitivity rapidly since 2020 [70]. Observing a gravitational wave at different places, we can get more detailed information about the origin.

Gravitational waves might give us information about stars or galaxies which cannot be obtained from observation by electromagnetic waves.

References

[1] Vanier J and Audoin C 1989 *The Quantum Physics Atomic Frequency Standards* (Bristol: Adam Hilger) p 610

[2] McCarcy D D and Seidelmann K P 2009 *Time: From Earth Rotation to Atomic Physics* (Weinheim: Wiley) pp 231–2

[3] Mungall A G 1971 *Metrologia* **7** 49

[4] Taylor J H and Weisberg J M 1982 *Astrophys. J.* **253** 908

[5] Hasegawa A *et al* 2004 *Metrologia* **41** 257

[6] Shimizu F 1970 *J. Chem. Phys.* **52** 3572

[7] Johns J W C and McKellar A R W 1973 *J. Mol. Spectrosc.* **48** 354

[8] Freund S M *et al* 1974 *J. Mol. Spectrosc.* **52** 38

[9] Job V A *et al* 1983 *J. Mol. Spectrosc.* **101** 48

[10] Adler F *et al* 2004 *Opt. Express* **12** 5872

[11] Haensch T W and Shawlow A L 1975 *Opt. Commun.* **13** 68

[12] Griffith W C *et al* 2009 *Phys. Rev. Lett.* **102** 101601

[13] Kozyryef I *et al* 2017 *Phys. Rev. Lett.* **119** 133002

[14] Metcalf H J and van der Straten P 1999 *Laser Cooling and Trapping* (Springer-Verlag New York, Inc.)

[15] Dalibard J and Cohen-Tannoudji C 1989 *J. Opt. Soc. Am.* B **6** 2023

[16] Morigi G *et al* 1999 *Phys. Rev.* A **59** 3797

[17] Hamann S E *et al* 1998 *Phys. Rev. Lett.* **80** 4149

[18] Ketterle W and van Druten N J 1996 *Adv. At. Mol. Opt. Phys.* **37** 181

[19] Kajita M 2022 *Ion Traps: A Gentle Introduction* (Bristol: IOP Publishing) 1–6

[20] Roos C F *et al* 2000 *Phys. Rev. Lett.* **85** 554

[21] Clairon A, Salomon C, Guellati and Phillips W 1991 *Europhys. Lett.* **16** 165

[22] Heavner T P *et al* 2005 *Metrologia* **42** 411

[23] Ovchinnikov Y and Marra G 2011 *Metrologia* **48** 87

[24] Chou C W *et al* 2010 *Phys. Rev. Lett.* **104** 070802
Brewer S M 2019 ArXiv: 1902.07694

[25] Huntemann N *et al* 2016 *Phys. Rev. Lett.* **116** 063001

[26] Ushijima I *et al* 2015 *Nat. Photonics* **9** 185

[27] Nicholson T L *et al* 2015 *Nat. Commun.* **6** 6896

[28] Nemitz N *et al* 2016 *Nat. Photonics* **10** 258

[29] Takano T *et al* 2016 *Nat. Photonics* **10** 662

[30] Takano Y and Katori H 2021 *J. Geod.* **95** 93

[31] Schioppo M *et al* 2022 *Nat. Commun.* **13** 212

[32] Knappe S *et al* 2005 *Opt. Lett.* **30** 2351

[33] Yano Y *et al* 2014 *Phys. Rev.* A **90** 013826

[34] Anderson P W 1949 *Phys. Rev.* **76** 647

[35] Landau L D and Lifshits E M 1970 *Quantum Mechanics* (Tokyo: Tosho) p 584 (in Japanese)

[36] Katori H, Kunigita H and Ido T 1995 *Phys. Rev.* A **52** R4324

[37] Kajita M 2004 *Phys. Rev.* A **69** 012709

[38] Anderson M H, Ensher J R, Matthews M R, Wieman C E and Cornell E A 1995 *Science* **269** 19
[39] Stamper-Kurn D M *et al* 1998 *Phys. Rev. Lett.* **80** 2027
[40] Hau L V, Harris S E, Dutton Z and Behroozi C H 1999 *Nature* **397** 594
[41] Donley E A *et al* 2001 *Nature* **412** 295
[42] MIT 1997 *MIT physicists create first atom laser* (MIT) http://web.mit.edu/newsoffice/1997/atom-0129.html (accessed 31 July 2006)
[43] Hara H *et al* 2014 *J. Phys. Soc. Jpn.* **83** 014003
[44] Joergens R *et al* 2008 *Nature* **455** 204
[45] Shu K *et al* 2017 *J. Phys. Conf. Ser.* **791** 012007
[46] Iijima H *et al* 2001 *J. Phys. Soc. Jpn.* **70** 3255
[47] Terada S *et al* 2013 *Nature* **500** 315
[48] Monroe C *et al* 1996 *Science* **272** 1131
[49] Shinjo A *et al* 2021 *Phys. Rev. Lett.* **126** 153604
[50] Haake F 2001 *Quantum Signatures of Chaos* (Berlin: Springer)
[51] Federal Aviation Administration 2014 Satellite Navigation Global Positioning System (GPS)
[52] Sicilia Ono B and Khatib O 2008 *Springer Handbook of Robotics* (Springer Science & Business Media)
[53] Zang L *et al* 2019 *Sensors* **19** 222
[54] Benioff P 1980 *J. Stat. Phys.* **22** 563
[55] Shor P W 1982 *SIAM J. Comput.* **26** 1484
[56] Deutsch D and Jozsa R 1992 *Proc. R. Soc. Lond. A: Math. Phys. Sci.* **439** 553
[57] Shor P W 1994 Proc.: 35th IEEE FOCS p 124
[58] Galderbank A R and Shor P W 1996 *Phys. Rev.* A **54** 1098
[59] Grover L K 1996 *Proc. of 28th Annual ACM Symp. on the theory of computing (May 1996)* 212
[60] Haroche S and Raimond J-M 1997 Le chat de Schroedinger se prete a l'experience *La Recherche* **301** 50
[61] Cirac J J and Zoller P 1995 *Phys. Rev. Lett.* **74** 4091
[62] DiVincenzo D P 1995 *Phys. Rev.* A **51** 1015
[63] Monroe C R *et al* 1995 *Phys. Rev. Lett.* **75** 4714
[64] Monroe C R and Wineland D 2008 *Quantum Computing with Ions* (Scientific American) https://scientificamerican.com/article/quantum-computing-with-ions
[65] Li J *et al* 2018 *IEEE Sens. J.* **18** 8198
[66] Dang H B 2010 *Appl. Phys. Lett.* **97** 151110
[67] Vengalattore M *et al* 2007 *Phy. Rev. Lett.* **98** 200801
[68] Abbott B P *et al* 2016 *Phys. Rev. Lett.* **116** 061102
[69] Abbott R *et al* 2021 *Astrophys. J. Lett.* **915** L5
[70] Takamori A *et al* 2023 *Earth Planets Space* **75** 98
[71] Kajita M 2018 *Measuring Time* (Bristol: IOP Publishing)
[72] Kajita M 2019 *Measurement, Uncertainty and Lasers* (Bristol: IOP Publishing)
[73] Kajita M 2020 *Cold Atoms and Molecules* (Bristol: IOP Publishing)

Chapter 3

Unsolved mysteries in particle physics

This chapter introduces the mysteries of elementary particles after the introduction of antiparticles, four kinds of interactions, and quarks. The relationship between particles and antiparticles has been described by the charge conjugation + mirror image (CP) symmetry, which must be violated to explain the particle-dominant Universe. The violation of CP-symmetry was observed with the unequal transition rate between particles and antiparticles of K-mesons. The violation of the time reversal (T) symmetry was also observed. The violation of CP-symmetry was theoretically derived by Kobayashi and Maskawa, but this theory is not enough to describe the particle predominant Universe. It is also a mystery that the violation of the CP-symmetry has been observed only with the weak interaction. From the theoretical prediction, the CP-symmetry can be violated also with the strong interaction. The search for the electron electric dipole moment (eEDM) has been performed to observe the violation of CP- and T-symmetries more in detail. The upper limit of eEDM was reduced to 4.1×10^{-30} e cm. On the other hand, charge conjugation + mirror image + time reversal (CPT) symmetry must be conserved to maintain the Lorenz invariance. With the previous experimental results, the CPT-symmetry is conserved. The idea to unify four kinds of interactions is also introduced. There is mystery of discrepancy between the estimations of proton radius using the transition frequencies of H atom (proton + electron) or muonic H atom (proton + μ^-– particle). The violation of the parity symmetric energy structure is expected to be detected also with optical isomers of chiral molecules, with which we can explain the unequal abundance of both isomers.

3.1 Particle and antiparticle

The Dirac equation (equation (1.6.3)) indicates the existence of the solutions with negative rest energy $-mc^2$. To solve this mystery with electrons, Dirac proposed a model that the vacuum is a sea of electrons with negative energy state [1]. The electron is a Fermion, and multi-electrons cannot exist in a single quantum state.

doi:10.1088/978-0-7503-6239-9ch3

Electrons with positive energy cannot decay to the negative energy state when all the quantum states with the negative energy are filled as shown in figure 3.1. If one electron in the vacuum is excited with an energy of $2m_ec^2$ to a positive rest energy state, a conventional electron is produced. The absence of a negative-mass electron is observed as a positive charge particle of the same mass as the electron. This notion of an antiparticle was confirmed by the discovery of the positron [2]. Pair production and pair annihilation of electron and positron were also observed. Antiprotons (negative charge) and antineutrons were also later discovered. However, the idea of the electron-sea is no longer used for the following reason. If we apply this idea for Boson particles (no limit to the number of particles in the same quantum state), all particles can be in the negative energy state. All Boson particles in the positive rest energy should decay to the negative energy state. The existence of the Boson particles in the positive rest energy indicates that no negative rest energy state exists for Boson particles. However, antiparticles have also been discovered with Boson particles. A new interpretation was provided by Feynman [3].

We can simply interpret the phase of particle waves like the rotation of the hands of a clock. With positive rest energy, the clock hand rotates clockwise. Then negative rest energy is interpreted as anti-clockwise rotation as shown in figure 3.2. We can imagine that clockwise and anti-clockwise rotations equally exist. Electromagnetic fields give energy shifts to particles, which is interpreted as the acceleration or deceleration force of the rotation of the clock hand. The direction of the energy shift is inverted by the conjugation of the electric charge. As the antiparticle has conjugated electric charge, the clockwise rotation of particle wave and anti-clockwise rotation of antiparticle wave are symmetric.

Particles and antiparticles are pairs produced from energy. With the collision between particle and antiparticle, both are pair annihilated and energy is produced,

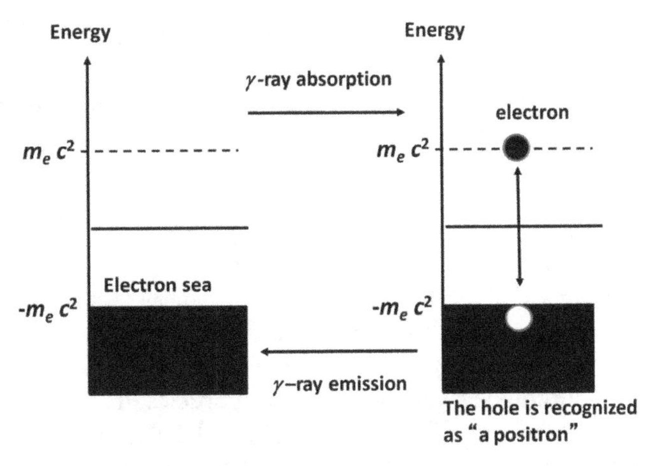

Figure 3.1. Dirac's 'electron-sea model' was later adapted to predict the existence of positrons. Reproduced from [54] and [55].

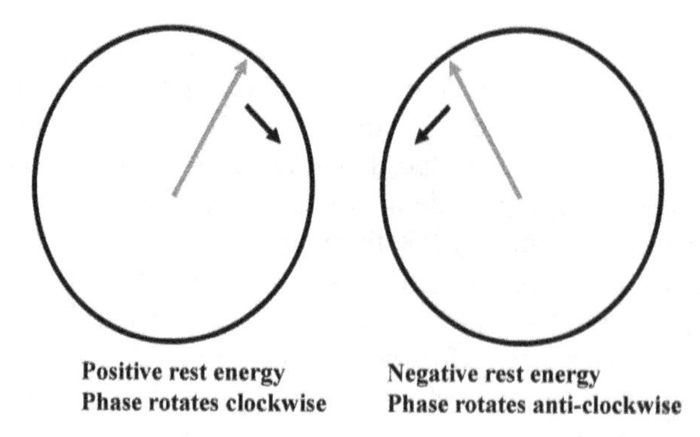

Figure 3.2. The image of the phase procedure of particle wave like a rotation of a clock hand. If we interpret the clockwise rotation with positive rest energy, negative rest energy can be interpreted as anti-clockwise rotation.

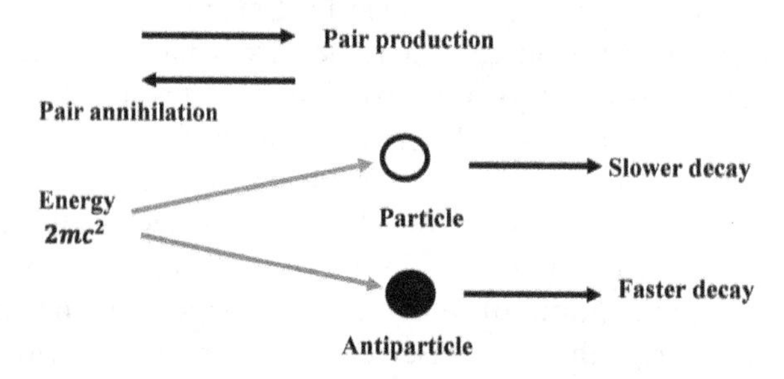

Figure 3.3. Pair production and annihilation of particles and antiparticles. Considering these procedures, numbers of particles and antiparticles must be equal. However, if the antiparticles decay faster than particles, the number of particles can be larger than that of antiparticles.

as shown in figure 3.3. Considering just these procedures, numbers of particles and antiparticles must be equal.

To describe the relation between particle and antiparticle, the following symmetries were proposed.

Charge (C) symmetry: all images of charge conjugation of real phenomena are equally possible as the real phenomena.

Parity (P) symmetry: all mirror images of real phenomena are equally possible with real phenomena.

Time (T) symmetry: all images of time reversal of real phenomena are equally possible as real phenomena.

If these symmetries strictly hold, the number of particles and antiparticles must be exactly equal. However, we don't see antiparticles in the Universe. With the current model, both particles and antiparticles existed soon after the birth of Universe. Antiparticles were annihilated via collision with particles. The number of particles exceeded that of antiparticles with the ratio of ($10^9 + 1$:10^9) and particles remain also after all antiparticles were annihilated. There must be some violation with these symmetries, for example, by the slight difference of decay time.

3.2 Research into radioactive rays

The research into radioactive rays provided important information about the structure of atoms and elementary particles. At first, x-rays (electromagnetic wave with frequency higher than ultraviolet area) were discovered by Roentgen in 1895 (before the discovery of the electron) [4]. In 1896, Becquerel discovered a natural radioactive ray from uranium. Its identity was clarified by Rutherford to be α-ray ($^4He^{2+}$: two protons and two neutrons), β-ray (electron), and γ-ray (electromagnetic waves with frequency higher than x-ray). The γ-ray was clarified not to be the flow of charged particles because its path was not bent by the magnetic field.

The radioactive rays are emitted with the change of the structure of nucleus. The size of the nucleus is five orders smaller than the atomic size, therefore, the quantum effects of the energy structure of the nucleus is much more significant (a larger energy gap) than that for the electronic energy structure of atoms. The energy of the α-ray is distributed in a narrow range. However, the energy of electrons emitted as the β-ray is continuously distributed in a wide range. Pauli gave a hypothesis that not only electrons, but also an unknown particle (named neutrino) is simultaneously emitted at the β-decay. The sum energy of electron and neutrino is constant [5]. In 1932, the neutron was discovered and the β-decay was clarified to be a transition from neutron n to (proton p + electron e^- + neutrino ν), which is possible because the mass (rest energy) of neutron is slightly larger than that of the proton.

The neutrino has no electric charge. The mass of the neutrino was assumed to be zero at first. In 1956, the neutrino was discovered observing the production of neutrons and positrons via the reaction between neutrinos and the nucleus in a water molecule [6]. There are three kinds of neutrinos: electron-neutrino (ν_e), μ-neutrino (ν_μ), and τ-neutrino (ν_τ). There are also antineutrinos: $\overline{\nu}_e$, $\overline{\nu}_\mu$, and $\overline{\nu}_\tau$. The spin is 1/2, but only the $M_S = -1/2$ state exists with neutrinos and only $M_S = 1/2$ state exists with antineutrinos. Therefore, C-symmetry is violated with neutrinos because the spin state is different. After distinguishing three kinds of neutrino and antineutrino, the β-decay is clarified to be $n \rightarrow p + e^- + \overline{\nu}_e$.

The direction of emitted electrons at the β-decay of Co atoms is localized in the direction antiparallel to the nuclear spin, as shown in figure 3.4 [7]. This result shows

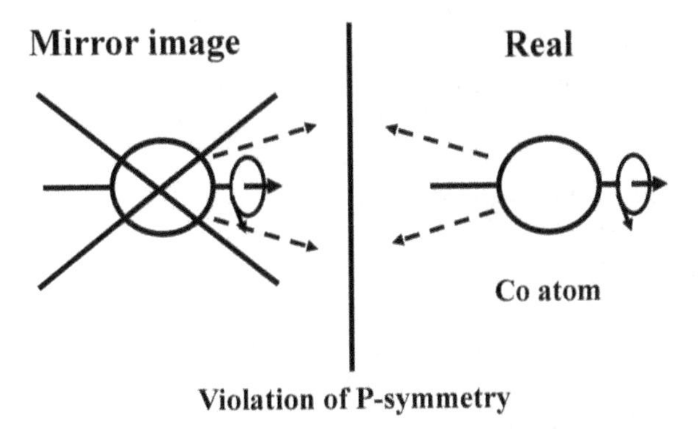

Figure 3.4. The emission of electron by the β-decay is localized in the inverse direction to the spin.

that P-symmetry is also violated at the β-decay. Section 3.3 introduces four interactions: strong interaction, electromagnetic interaction, weak interaction, and gravitational interaction. The β-decay is induced by the weak interaction, which violates the P-symmetry and provides a large role in making an inhomogeneous abundance of particles and antiparticles.

3.3 Particles in the first generation and the interactions

All forces are given by the exchange of particles. For example, electromagnetic force is given by the exchange of photons.

The nucleus is constructed by protons and neutrons, which are attracted by the nuclear force. The nuclear force potential given by the exchange of a particle X having a mass of m is given by a formula (called the Yukawa potential):

$$P_Y(r) \propto \frac{\exp\left(-\frac{r}{\lambda_Y}\right)}{r}$$

$$\lambda_Y = v_X \tau'_X$$

$$\tau'_X = \frac{\tau_X}{\sqrt{1-(v_X/c)^2}}, \tag{3.3.1}$$

where v_X and τ_X denote the mean velocity and decay time of X, respectively. The decay time in the nucleus should be considered with τ'_X taking the relativistic effect with the motion of X into account. For example, $1/\lambda_Y = 0$ when the mass of X is zero because it can move only at the speed of light. Using the relativistic definition of the momentum p_X and the de Broglie wavelength $\lambda_{dB}(=h/p_X)$, the following relation is derived.

$$\lambda_Y = \frac{p_X \tau_X}{m} = \frac{h\tau_X}{m\lambda_{dB}}. \tag{3.3.2}$$

Assuming $\lambda_{dB}/\tau_X = c$ ($\tau_X > \lambda_{dB}/c$ is required for the decay of whole wave broadening), λ_Y corresponds to the Compton wavelength estimated by $mc^2 = cp_X$. λ_Y is predicted from the size of the nuclear. Yukawa predicted the existence of a particle with a mass of the order of $0.11m_p$, where m_p is the mass of proton (1.6×10^{-27} kg, 938 MeV c^{-2}). Three kinds of π-mesons with the electric charge of ±e and neutral (π^+, π^-, and π^0) were later discovered. π^- is antiparticle of π^+. The antiparticle of π^0 is itself.

The mass and lifetime of π-mesons are

π^+ 0.149 m_p 2.6×10^{-8} s

π^0 0.144 m_p 8.5×10^{-17} s.

The nuclear force is not interpreted as a fundamental force, because π-mesons are not elementary particles (cannot be divided further), as shown below.

There were mysteries which could not be solved while considering proton p and neutron n as elemental particles; e.g., a neutron has non-zero magnetic moment despite neutral electric charge. Gell-Mann suggested the hypothesis that protons and neutrons are constructed by binding the elemental particles, called quarks [8]. There are up quarks (u) with the electric charge of +2e/3 and down quarks (d) with the electric charge of −e/3. As shown in figure 3.5, the proton is constructed by two up quarks and one down quark, while the neutron is constructed by one up quark and two down quarks. There are antiparticles for both quarks; $\bar{u}$ and $\bar{d}$. The masses of up and down quarks are of the order of 0.002 m_p and 0.005 m_p, respectively. The mass of protons and neutrons are two orders larger than the total mass of constructing quarks and are dominated by the binding energy between quarks. The force to bind quarks is called the 'strong interaction', which is given by exchanging eight kinds of gluons. Gluons are Boson particles (spin 1) without mass or electric charge. The β-decay is interpreted as $d \rightarrow u + e^{-} + \overline{\nu}_e$.

The strong interaction is described by the interaction between color charges in analogy with the three original colors of light. Quarks have color charges of red, blue, and green. Quarks tend to bind each other so that the total mixture of colors becomes white, just as electric charge becomes neutral by the bunch of positive and negative charges. The color charge of proton and neutron are white by the mixture of three original colors as shown in figure 3.5. Previously quarks have been discovered only in the bunched state with the total color of white because the strong interaction between quarks becomes stronger with larger distance.

It is a mystery how the spin of a proton or neutron is determined, although it has been clarified that the contribution of quarks is less than 30%, and the role of spin and orbital angular momentum of gluons is rather significant.

Antiquarks have charges of complementary colors. The total color can be white by the bunching of one quark and one antiquark. Pie-mesons are constructed as

$\pi^{+} = (u\bar{d})\ \pi^{-} = (\bar{u}d)\ \pi^{0} = \frac{(u\bar{u}) - (d\bar{d})}{\sqrt{2}}$.

The lifetime of π^0 is much shorter than $\pi^{\pm}$ because of pair annihilation between quarks and antiquarks.

There are no phenomena requiring the model that electron and neutrino are the binding of smaller particles. Therefore, they are elemental particles called 'leptons'.

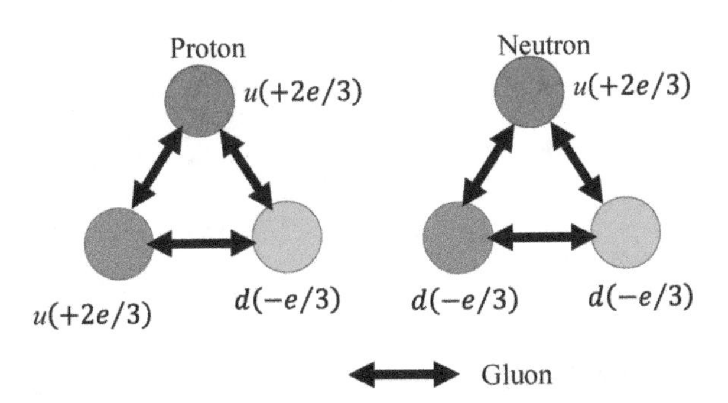

Figure 3.5. Protons and neutrons are constructed by up (u: electric charge +2e/3) and down (d: electric charge −e/3) quarks. The binding is caused between three color charges (red, blue, and green) and the mixture color becomes white.

The P-symmetry holds with the strong interaction and the electromagnetic interaction. The ratio of the electromagnetic force to the strong interaction is of the order of the fine structure constant $\alpha = e^2/2\varepsilon_0 hc = 0.007\ 297$.

The 'weak interaction' is several orders weaker than the strong interaction and the electromagnetic interaction. It is induced by the exchange of W^+ (electric charge +e), W^- (electric charge −e), or Z (electric neutral) Boson particles. The W^- Boson is the antiparticle of the W^+ Boson. The weak interaction works in a very small area (one order smaller than the proton radius 9×10^{-16} m) because the Boson particles have large mass (W Boson 80 m_p and Z Boson 90 m_p) and the motion range before decay (the lifetime is of the order of 10^{-25} s) is very small. W and Z particles were experimentally discovered in 1983 [9]. The weak interaction has a very special characteristic because it can violate the C- and P-symmetries. It can induce the transition from the down quark to the up quark. The β-decay is interpreted with the following two procedures:

$$d \rightarrow u + W^-$$

$$W^- \rightarrow e^- + \overline{\nu}_e.$$

Gravitational interaction is the distortion of the space, which propagates as a wave with the speed of light (see section 1.4). However, it is described also as a flow of a particle named the 'graviton' just like light is an electromagnetic wave and also a flow of photons. The mass of the graviton is zero and it works at long distance as shown by equation (3.3.1) taking $1/\lambda_Y = 0$. The graviton has never been discovered. The gravitational force is 40 orders weaker than the strong interaction; this order of magnitude of weakness is also a mystery of modern physics.

3.4 Particles in the higher generation

There are μ-particle (μ^-) and τ-particle (τ^-), which have almost the same property as electrons (e^-) except for the larger mass and shorter lifetime. The mass of μ^- and τ^- are $250m_e$ and $3550m_e$ (m_e: electron mass $m_e/m_p = \frac{1}{1840}$), respectively. Both particles decay with

$$\mu^- \rightarrow e^- + \overline{\nu}_e + \nu_\mu \text{ life time } 2.2 \times 10^{-6}\ \text{s}$$

$$\tau^- \rightarrow \pi^- + \pi^0 + \nu_\tau, \quad 25.5\%$$

$$\rightarrow e^- + \overline{\nu}_e + \nu_\tau, \quad 17.8\%$$

$$\rightarrow \mu^- + \overline{\nu}_\mu + \nu_\tau, \ 17.4\% \text{ lifetime } 2.9 \times 10^{-13}\text{s}.$$

We consider μ^- and τ^- as the second and third generation of electron, as shown in table 3.1. Particles in the higher generation decay and transform to the lower generation. There are also antiparticles of electron, μ-particle and τ-particle; e^+, μ^+, and τ^+.

The electron-neutrino (ν_e), μ-neutrino (ν_μ), and τ-neutrino (ν_τ) are the neutrinos in the first, second, and third generation. The periodic change of the abundance ratio of

Table 3.1. List of particles from the first to third generations.

	Charge	Generation 1	Generation 2	Generation 3
Quark	$+2e^+/3$	Up quark (u)	Charm quark (c)	Top quark (t)
	$-e^-/3$	Down quark (d)	Strange quark (s)	Bottom quark (b)
Lepton	$-e$	Electron (e^-)	μ-Particle (μ^-)	τ-Particle (τ^-)
	0	Electron-neutrino (ν_e)	μ-Neutrino (ν_μ)	τ-Neutrino (ν_τ)

Table 3.2. List of mesons, which are discussed in this book.

Meson	Antimeson	Structure	Lifetime (s)
π^+	π^-	$u\bar{d}$	2.6×10^{-8}
π^0	π^0	$(u\bar{u} - d\bar{d})/\sqrt{2}$	8.4×10^{-17}
K^+	K^-	$u\bar{s}$	1.2×10^{-8}
K_S^0	K_S^0	$(d\bar{s} + \bar{d}s)/\sqrt{2}$	8.9×10^{-11}
K_L^0	K_L^0	$(d\bar{s} - \bar{d}s)/\sqrt{2}$	5.2×10^{-8}
B^+	B^-	$u\bar{b}$	1.7×10^{-12}
B_0	$\overline{B_0}$	$d\bar{b}$	1.5×10^{-12}

three neutrinos (neutrino oscillation) were theoretically analyzed [10]. However, the observation of neutrino oscillation should not be observed if the mass of neutrinos were zero and move with the speed of light, because the theory of relativity prohibits any temporal change in a frame moving with the speed of light. The neutrino oscillation was observed by different research groups and this result indicates that the mass of neutrino is non-zero [11].

Three generation particles exist also for quarks, as shown in table 3.1. At the second generation, the charm quark (c) with a mass of 1.3 m_p and the strange quark (s) with a mass of 0.11 m_p exist. The top quark (t) with a mass of 183 m_p and the bottom quark (b) with a mass of 4.5 m_p are in the third generation. Quarks are difficult to discover as single isolated particles (section 3.3) and their existence was recognized by discoveries of different mesons. Some examples of mesons are listed in table 3.2. Here, K_S^0 and K_L^0 are symmetrical and antisymmetrical coupling between K^0 ($d\bar{s}$) and $\overline{K^0}$ ($\bar{d}s$).

Hyper-nuclei include hyperons, which include higher generation quarks. Table 3.3 lists Λ-particles, which are hyperons constructed by u, d, and higher generation quarks [12]. The precision measurement of binding energy of the hyper-nucleus gives important information of nuclear force. Λ^0 decays to $p + \pi^-$ or $n + \pi^0$. The lifetime was theoretically estimated to be of the order of 10^{-23} s, but it is much longer as shown in table 3.3. This discrepancy was explained by introducing a quantum number 'strangeness'. Strangeness (given by the number difference of s and $\bar{s}$) is conserved with the strong interaction and electromagnetic interaction and the

Table 3.3. List of Λ-particles. Λ_t^+ has not been discovered yet, although its existence was predicted [12].

Particles	Structure	Mass (m_p)	Electric charge (e)	Lifetime (s)
Λ^0	uds	1.19	0	2.6×10^{-10}
Λ_c^+	udc	2.44	+1	2.0×10^{-13}
Λ_b^0	udb	5.99	0	1.4×10^{-12}
Λ_t^+	udt		+1	

Table 3.4. List of Ξ-particles with the lifetime longer than 10^{-12} s [13].

Particles	Structure	Mass (m_p)	Electric charge (e)	Lifetime (s)
Ξ^0	uss	1.40	0	2.9×10^{-10}
Ξ^-	dss	1.41	−1	1.6×10^{-10}
Ξ_b^0	usb	6.17	0	1.4×10^{-12}
Ξ_b^-	dsb	6.17	−1	1.4×10^{-12}

decay rate is suppressed. The lifetime of Λ^0 is limited by the weak interaction, which can change the strangeness. Λ-particles slide into the center of the hyper-nucleus, and it binds the protons and neutrons more tightly due to its stronger nuclear force than a conventional nucleus. For example, the ^{7}Be nucleus is 17% smaller. There is also Σ^+-particle with the structure of uus (lifetime 8×10^{-11} s). There are also Ξ-particles including two high generation quarks, which are listed in table 3.4 [13]. The lifetimes of Ξ^0 and Ξ^- are of the same order as that of Λ^0, because they are limited by the decay of s induced by the weak interaction.

There might be a question of why there is no fourth generation. Three kinds of neutrino have been discovered, but no fourth generation has been discovered. The abundance ratio between proton and neutron is 6:1 (estimated from the ratio of numbers of H and He atoms), which is consistent with the model with three generations. However, we cannot perfectly rule out the possibility of the existence of fourth-generation particles. We can imagine that also with very large mass and very short lifetime if fourth-generation particles exist they will be difficult to discover.

3.5 Violation of the CP-symmetry and T-symmetry

As shown in section 3.2, the C- and P-symmetries are violated with the weak interaction. However, CP-symmetry (charge conjugate + mirror reflection) was expected to hold, because $\overline{\nu}_e$ is the mirror image of ν_e. The CP-symmetry between the $\mu^- \rightarrow e^- + \overline{\nu}_e + \nu_\mu$ and $\mu^+ \rightarrow e^+ + \nu_e + \overline{\nu}_\mu$ transitions hold as shown in figure 3.6, with which the C and P-symmetries are violated.

The violation of the CP-symmetry was discovered by observing the mixture of particles and antiparticles of the K^0 meson (K_S^0 and K_L^0). The lifetime of K_S^0 is much shorter than K_L^0. Therefore, pure K_L^0 should be attained after the lifetime of K_S^0. But

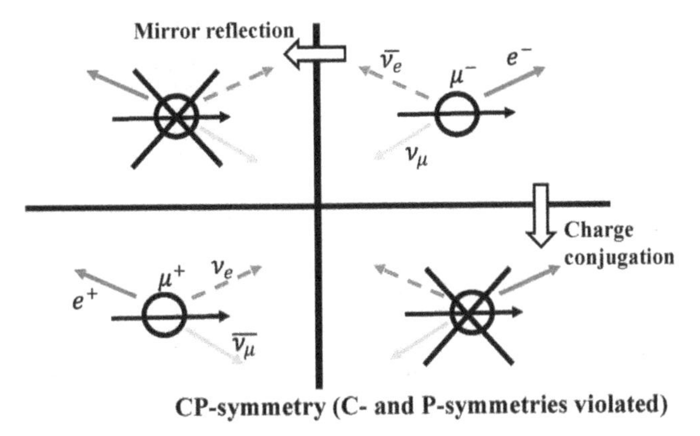

Figure 3.6. The CP-symmetry holds with the decay of $\mu^{\pm}$-particles, although C- and P-symmetries are violated.

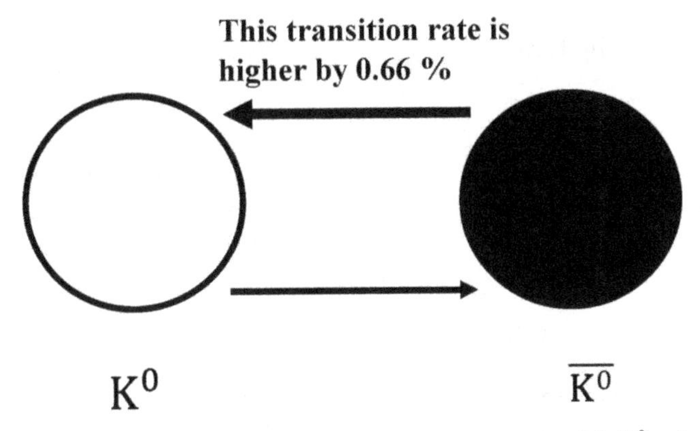

Figure 3.7. Violation of the T-symmetry with the transformation between particle K^0 and antiparticle $\overline{K^0}$ of kaon.

K_S^0 was observed also after a period much longer its lifetime [14]. This result indicates that the $K_L^0 \to K_S^0$ transition (change of symmetry) exists. The change of symmetry of an electric neutral particle indicates the violation of the CP-symmetry. The violation of the T-symmetry was also indicated from the experimental result showing that the $K^0 \to \overline{K^0}$ transition rate is higher than the $\overline{K^0} \to K^0$ transitions rate with the ratio of 0.66% [15] as shown in figure 3.7.

The difference of lifetime between B_0 and $\overline{B}_0$ with the order of 10% was experimentally confirmed, as shown in figure 3.8 [16]. This result indicates the violation of the T-symmetry. The non-equality of numbers of particles and antiparticles is derived from the difference of the lifetime as shown in figure 3.3.

There is currently a project to compare the neutrino oscillation between ν and $\bar{\nu}$, which might indicate the violation of the CP- and T-symmetries [17]. Previous experimental results indicate a slight difference between the abundance ratio of ν_e and $\overline{\nu}_e$, but this difference is not large enough in comparison with current measurement uncertainty. A higher sensitivity of the neutrino signal is required to confirm with the lower statistic measurement uncertainty.

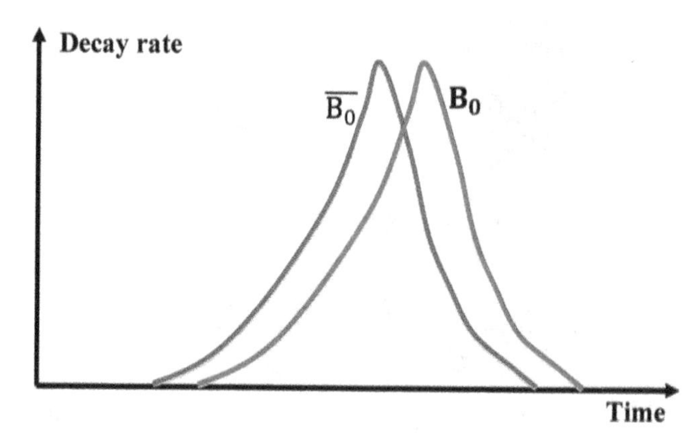

Figure 3.8. A rough description of the decay rate of B_0 and $\overline{B_0}$, which indicates the violation of the T-symmetry [16].

3.6 Theoretical derivation of the violation of CP-symmetry

The theoretical derivation of the CP-symmetry was given from the coupling of quarks with three generations as follows.

The transition between quarks induced by the weak interaction was well described by treating quarks as the mixture of different generations. This mixture is induced by the strong interaction. In 1963, Cabibbo proposed the Cabibbo angle θ_C with which the β-decay is given by the d′ → u transition, where

$$\langle \mathrm{d}' \rangle = \cos\theta_C \langle \mathrm{d} \rangle + \sin\theta_C \langle \mathrm{s} \rangle. \tag{3.6.1}$$

Taking $\theta_C = 0.0724\pi$, the experimental result was described well. After the discovery of the charm quark (c), the s′ → c transition was also observed with

$$\langle \mathrm{s}' \rangle = -\sin\theta_C \langle \mathrm{d} \rangle + \cos\theta_C \langle \mathrm{s} \rangle. \tag{3.6.2}$$

From these relations, the Cabibbo matrix was proposed to give the mixture between two generations causing the d′ → u and s′ → c transitions:

$$\begin{pmatrix} \langle \mathrm{d}' \rangle \\ \langle \mathrm{s}' \rangle \end{pmatrix} = \begin{bmatrix} \cos\theta_C & \sin\theta_C \\ -\sin\theta_C & \cos\theta_C \end{bmatrix} \begin{pmatrix} \langle \mathrm{d} \rangle \\ \langle \mathrm{s} \rangle \end{pmatrix}. \tag{3.6.3}$$

The Cabibbo matrix is described using only one real parameter. Describing the coupling coefficients using complex numbers, equation (3.6.1) is rewritten as

$$\langle \mathrm{d}' \rangle = \cos\theta_C \langle \mathrm{d} \rangle + \exp\left(i\delta_p\right) \sin\theta_C \langle \mathrm{s} \rangle. \tag{3.6.4}$$

The effect of $\exp\left(i\delta_p\right)$ is interpreted as the phase shift between wavefunctions $\langle \mathrm{d} \rangle$ $(\propto \exp(2\pi i \nu_\mathrm{d} t))$ and $\langle \mathrm{s} \rangle$ $\left(\propto \exp\left(2\pi i \nu_\mathrm{s} t + i\delta_p\right)\right)$. Just shifting the origin of time $t \to t' = t + \delta_p/(\nu_\mathrm{d} - \nu_\mathrm{s})$, the wavefunctions are given by $\langle \mathrm{d} \rangle$ $(\propto \exp(2\pi i \nu_\mathrm{d} t'))$ and $\langle \mathrm{s} \rangle$ $(\propto \exp(2\pi i \nu_\mathrm{s} t'))$ and the effect of $\exp\left(i\delta_p\right)$ is eliminated. While considering two generations of quarks, the violation in the CP-symmetry is not derived because the value of θ_C is also the same for the $\overline{d'} \to \overline{u}$ and $\overline{s'} \to \overline{c}$ transitions.

However, considering the coupling between three states, the effect of the phase shift cannot be eliminated. Kobayashi and Maskawa indicated that the coupling between quarks in the three generations are given by complex numbers [18]. The phase of wavefunctions of particles and antiparticles are inversed (clockwise or anti-clockwise rotation of clock hand shown in section 3.1). Therefore, the coupling coefficients for antiparticles are complex conjugate of those for particles. The coupling coefficient is given by the Cabibbo–Kobayashi–Maskawa (CKM) matrix

$$\begin{pmatrix} \langle d' \rangle \\ \langle s' \rangle \\ \langle b' \rangle \end{pmatrix} = \begin{bmatrix} 1 - \lambda^2/2 & \lambda & A\lambda^3(\rho - i\eta) \\ -\lambda & 1 - \lambda^2/2 & A\lambda^2 \\ A\lambda^3(1 - \rho - i\eta) & -A\lambda^2 & 1 \end{bmatrix} \begin{pmatrix} \langle d \rangle \\ \langle s \rangle \\ \langle b \rangle \end{pmatrix}$$

$$\lambda = 0.2257,\ A = 0.814,\ \rho = 0.135,\ \eta = 0.349. \tag{3.6.5}$$

This theory was not accepted immediately after the publication, because only three kinds of quarks (u, d, and s) were discovered at that time. It took 21 years until all other quarks (c, b, and t) were discovered, because very high energy is required to produce them. The difference of lifetimes between B_0 and $\overline{B}_0$ was estimated with this theory, which is in good agreement with experimental result [16]. The Kobayashi–Maskawa theory derived the non-equal numbers of particles and antiparticles. However, it is not enough to derive the number of particles in the Universe. The ratio of the number of particles to that of photons estimated by Kobayashi–Maskawa theory is ten orders smaller than the result of observation.

3.7 Strong CP-problem

Regarding the strong interaction, there is a mystery called the 'strong CP-problem' [19]. With the strong interaction by the color charge, there is a parameter called the 'vacuum angle', which cannot be determined uniquely. With the non-zero vacuum angle, the CP-symmetry is violated. However, the conservation of the CP-symmetry has been confirmed experimentally, taking the upper limit of the vacuum angle 10^{-9}. It is a mystery why the vacuum angle is so small. To solve this problem, the existence of a spinless particle (axion) with small mass was predicted [20]. The color charge (strong interaction) produces an effective periodic potential and the axion has an oscillation around the position where the potential is minimum (zero vacuum angle). The axion is one candidate for dark matter (section 4.3). Many researchers are eager to discover the axion, as shown in chapter 4, because it may solve the mystery about the unequal abundance of particles and antiparticles from the small but non-zero oscillation amplitude of vacuum angle.

3.8 Search of the electron electric dipole moment

Recently, the search for an electric dipole moment (EDM) in the electron (eEDM) has drawn interest as direct evidence of T-symmetry violation. If the value of eEDM

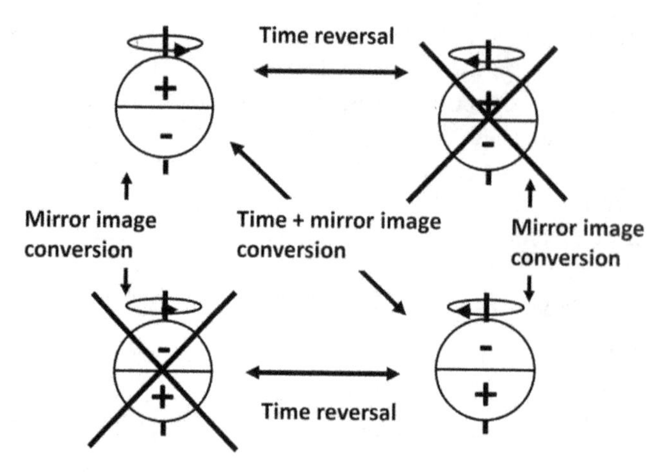

Figure 3.9. Existence of the electron electric dipole moment (eEDM) as evidence for the violation of T- and P-symmetries. Reproduced from [54].

is measured to be much larger than the value estimated by Kobayashi–Maskawa theory (10^{-38} e cm), it might give a consistent estimation of the abundance ratio between particles and antiparticles.

If eEDM exists, there are at least two states $\pm d_e$ with distinct directions (figure 3.9). The electron has two spin states $M_S = \pm 1/2$, and there are just two states for the electron. If the electron spin and eEDM are independent states, more than four electrons could exist in a same state given by the principal quantum number n, rotational quantum number L, and magnetic quantum number M_L. The possible eEDM state can be one state corresponding to the spin state: the possible two-electron states are ($M_S = 1/2$, $+d_e$) and ($M_S = -1/2$, $-d_e$). Time reversal gives transformations ($M_S = 1/2$, $+d_e$) → ($M_S = -1/2$, $+d_e$) and ($M_S = -1/2$, $-d_e$) → ($M_S = 1/2$, $-d_e$), which are not possible; therefore, T-symmetry is violated. The P-symmetry is also violated as the mirror image conversion gives the transformation ($M_S = 1/2$, $+d_e$) → ($M_S = 1/2$, $-d_e$). The PT-symmetry is conserved also if eEDM exists.

The search for the eEDM should be done by measuring the energy shift induced by the electric field E_e. This is not possible, however, with free electrons, which are accelerated by the electric field. Therefore, atoms or molecules having non-zero electron spin are used for the measurement (figure 3.10). When the magnetic and electric fields are zero, there is no energy difference between ($M_S = 1/2$, $+d_e$) and ($M_S = -1/2$, $-d_e$) states. A magnetic field is applied to fix the direction of the electron spin and eEDM. Given a magnetic field B, there are two Zeeman energy shifts with coefficients $\pm a_Z M_S$ and the transition frequency between both states is

$$\nu_Z = 2\frac{a_Z B}{h}. \tag{3.8.1}$$

The coefficient a_Z is given by the relation between electron spin and the electron orbital angular momentum (without electron orbital angular momentum

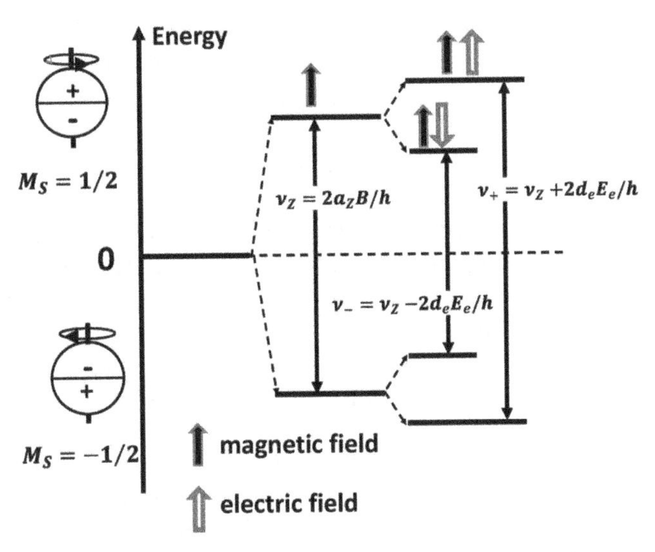

Figure 3.10. Measurement of eEDM in atoms or molecules having non-zero electron spin which provide magnetic and electric fields, simultaneously (parallel or antiparallel).

$a_Z/h = 1.4$ MHz G^{-1}). When an electric field E_e is also applied parallel or antiparallel to the magnetic field, a Stark energy shift appears and the transition frequency $\nu_\pm$ is given by

$$\nu_\pm = \nu_Z \pm \frac{2d_e E_e}{h}$$

$\nu_+(\nu_-)$: with electric field parallel (antiparallel) to the magnetic field. (3.8.2)

Measuring the difference between these two transition frequencies,

$$\nu_+ - \nu_- = \frac{4d_e E_e}{h} \tag{3.8.3}$$

yields a value for the eEDM.

With the value of d_e estimated from the Kobayashi–Maskawa theory (10^{-38} e cm) [18], the effect of eEDM shown in equation (3.8.3) is not detectable. However, the CP-symmetry violation determined by this theory is not enough to derive a particle-dominant Universe and a range of values of 10^{-26}–10^{-30} e cm is estimated by different theories. The detection with electric field intensities of 10^5 V cm^{-1} (highest electric field artificially generated) is not realistic, because $\nu_+ - \nu_- < 10^{-7}$ Hz. However, electrons in atoms or molecules have the possibility to experience an internal electric field with the order 10^{10}–10^{11} V cm^{-1} as shown below. The average electric field applied to electrons in atoms or molecules is zero in the laboratory frame, otherwise they would be accelerated. However, electrons moving with high velocity also experience a magnetic field from relativistic effects and the forces from the internal electric field and magnetic field are balanced. Therefore, electrons in atoms or molecules having heavy nuclei can experience internal electric fields higher than 10^{10} V cm^{-1}. The direction of the internal electric field is random, but it can be aligned using an external electric field (figure 3.11). We interpret that the applied external electric field is amplified with several orders by the alignment of internal electric field.

For the search of eEDM, high internal electric field and the precision frequency measurement are required. The internal electric field is high with atoms or molecules having a heavy nucleus. To align the direction of the internal electric field, molecules with large EDM are advantageous. For the precision measurement, it is preferable to use cold atoms or molecules. Cold polar molecules having heavy nuclei are most advantageous. But it was difficult to obtain cold molecules until 2010 and there was a choice between measurement with ultra-cold heavy atoms (laser cooled) or thermal polar molecules. Recently measurement with cold polar molecules has been performed as shown below.

In 2011, a group at Imperial College London obtained an upper limit for eEDM of 1.05×10^{-27} e cm using YbF molecules in a thermal beam [21]. In 2014, a joint group from Harvard and Yale obtained an upper limit for eEDM of 0.87×10^{-28} e cm from a measurement using ThO molecules cooled by the vapor of liquid He [22]. In 2018, this upper limit was reduced to 1.1×10^{-29} e cm [23]. This group is revising the apparatus to reduce the measurement uncertainty by one order. A group in the Joint Institute of University of Colorado Boulder and NIST (JILA) attained the upper limit of 4.1×10^{-30} e cm using trapped HfF^+ molecular ion [24]. Recently, the group at Imperial College, London, succeeded in decelerating YbF molecules by laser cooling [25], The search of eEDM might be possible using laser cooled molecules. The Harvard group is now developing a laser cooling system to decelerate the SrOH molecule [26].

T-symmetry violation would also be demonstrated by the presence of EDMs in other particles. There is the possibility to search for an EDM in a nucleus for which

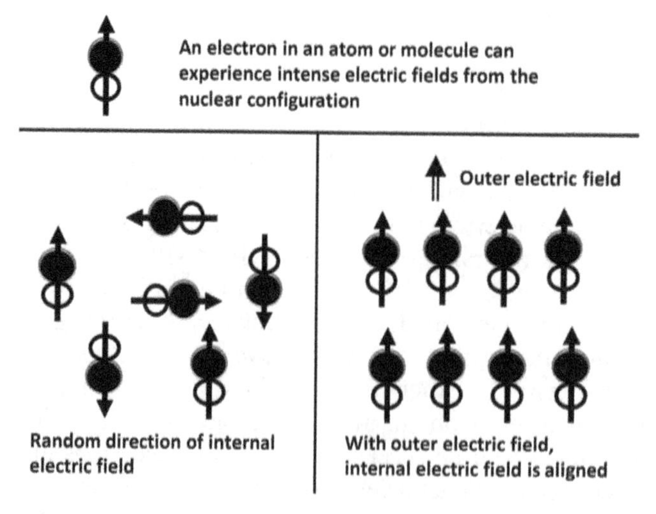

Figure 3.11. Alignment of the internal electric field. Reproduced from [54].

the internal electric charge distribution is localized (Schiff moment). The upper limits for the ^{199}Hg and ^{129}Xe nuclei are estimated to be 3.1×10^{-29} e cm [27] and 7×10^{-27} e cm [28], respectively.

3.9 CPT-symmetry

When the CP-symmetry is violated, the T-symmetry is also violated as shown in section 3.5. And the CPT-symmetry (charge conjugation + mirror reflection + time reversal) is conserved with the decay of the K_0 meson.

To guarantee the Lorenz invariance (all physical laws in a coordinate must hold with any coordinates given by the Lorenz transform), the CPT-symmetry must conserve strictly [29]. A simple interpretation of the CPT-symmetry is shown using the Dirac equation (1.6.3)

$$\left(\frac{h}{2\pi i}\frac{\partial}{\partial t} - q_e\Phi_{el}\right) = \sum_{Q=x,y,z}\begin{pmatrix} 0 & \sigma_Q \\ \sigma_Q & 0 \end{pmatrix}\left(\frac{h}{2\pi i}\frac{\partial}{\partial Q} + q_e A_Q\right) + \begin{pmatrix} I & 0 \\ 0 & -I \end{pmatrix} mc^2$$

σ_Q: Pauli matrix $Q = x, y, z$

A_Q: components of magnetic field vector potential in the Q – direction

Φ_{el}: electric voltage. (3.9.1)

Giving the CPT-transform ($q_e \to -q_e$, $Q \to -Q$, $t \to -t$) to equation (3.9.1),

$$\left(\frac{h}{2\pi i}\frac{\partial}{\partial t} - q_e\Phi_{el}\right) = \sum_{Q=x,y,z}\begin{pmatrix} 0 & \sigma_Q \\ \sigma_Q & 0 \end{pmatrix}\left(\frac{h}{2\pi i}\frac{\partial}{\partial Q} + q_e A_Q\right) + \begin{pmatrix} -I & 0 \\ 0 & I \end{pmatrix} mc^2 \quad (3.9.2)$$

is obtained. Equation (3.9.2) is also valid transforming the solution of wavefunction as the four-dimensional vector by $\begin{pmatrix} \vec{u} \\ \vec{w} \end{pmatrix} \to \begin{pmatrix} \vec{w} \\ \vec{u} \end{pmatrix}$, where $\vec{u}$ and $\vec{w}$ are two-dimensional vectors.

To conserve the CPT-symmetry, the equalities of the absolute value of the electric charge and the mass are required between the particles and the antiparticles. Some experiments to confirm the CPT-symmetry are introduced below.

3.9.1 Precision measurement of energy structure of antiproton helium

The equality of (electric charge)^2 $\times$ (mass) between proton and antiproton was confirmed from precision measurements of the transition frequency of the antiproton helium atom (figure 3.12) [30]. An antiproton helium atom consists of a helium nucleus (charge +2), one electron (charge −1), and one antiproton (charge −1). The orbital radius of the antiproton is much less than that of an electron. The electron cloud has a role to prevent other particles colliding with the antiproton. There is no electric interaction between the antiproton and the electron. The energy level of the antiproton determined by the interaction of the He^{2+} nucleus is given by

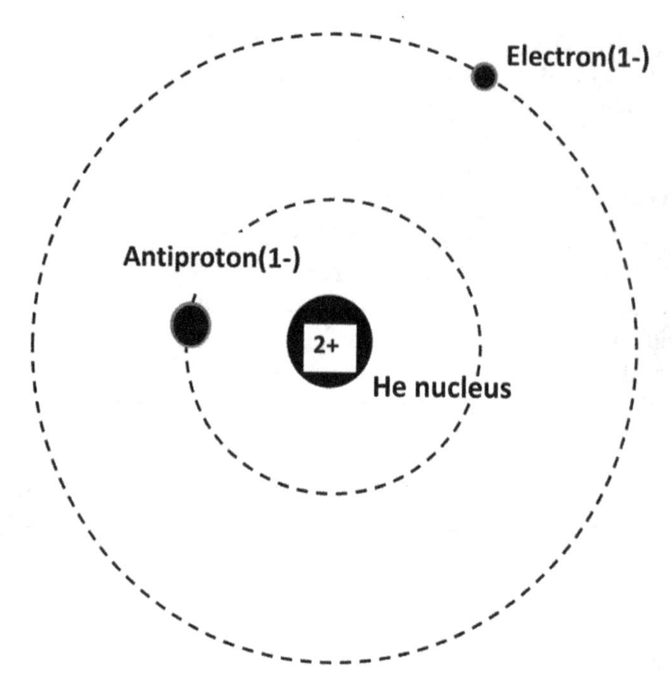

Figure 3.12. Schematic depicting the structure of an antiproton helium atom, consisting of a He nucleus, antiproton, and electron. Reproduced from [54]. Copyright IOP Publishing Ltd. All rights reserved.

$$E_{\bar{p}} = -\frac{\mu_{\bar{p}} e'^2 e^2}{2h^2 \varepsilon_0^2} \frac{1}{n^2}$$

$\mu_{\bar{p}} = \frac{m_{\bar{p}} m_{He^{2+}}}{m_{\bar{p}} + m_{He^{2+}}}$ $m_{\bar{p}}$, $m_{He^{2+}}$: mass of antiproton and He^{2+} nucleus

$$e'\text{: charge of anti} - \text{proton.} \tag{3.9.3}$$

From precision measurements of the transition frequencies, the ratio between $m_{\bar{p}}\ e'^2$ and $m_e e^2$ was obtained. After (m_p/m_e) was determined, the equality between $m_{\bar{p}}\ e'^2$ and $m_p e^2$ was confirmed within a fractional uncertainty of 10^{-9}.

3.9.2 Comparison of the frequency of cyclotron motion

Charged particles (charge q_e) with the mass of m have a circular motion when there is a magnetic field B with the frequency of

$$\nu_{cyc} = \frac{q_e B}{2\pi m}. \tag{3.9.4}$$

Comparing the cyclotron frequencies of proton and antiproton with the magnetic field of 1.95 T, the equality of $|e/m_p|$ and $|e'/m_{\bar{p}}|$ was confirmed with the uncertainty of 1.6×10^{-11} [31].

3.9.3 Comparison of the transition frequencies of atom and anti-atom

The equality of the ($n = 1$, $L = 0$) − ($n = 2$, $L = 0$) transition frequencies (n: principal quantum number, L: rotational quantum number) between an H atom and an anti-H atom (consisting of an antiproton and positron) was also confirmed (figure 3.13) with an uncertainty of 2×10^{-12} [32]. The ($n = 1$, $L = 0$) − ($n = 2$, $L = 1$) transition frequencies of H and anti-H atoms were also compared, and they were in good agreement within the uncertainty of 5×10^{-8} [33]. The statistical measurement uncertainty (Appendix C) of the $\Delta L = 1$ transition is higher than that for the $\Delta L = 0$ transition because the spectrum linewidth determined by the spontaneous emission rate is larger by six orders.

For the precise measurement of the transition frequency of anti-H atoms, long observation time and low kinetic energy were required. Laser cooling of the anti-H atom was recently demonstrated [34], therefore, the further reduction of the measurement uncertainty is expected. We will see if the CPT-symmetry is guaranteed with lower uncertainty, or its violation is discovered (if so, new physics is required).

3.10 The identity of the neutrino

The wavefunction of antiparticles are complex conjugates of those of particles. Section 3.1 indicates that the phase of the particles rotates clockwise while that of antiparticle rotates anti-clockwise. Majorana suggested in 1937 that electrically neutral particles with a spin of 1/2 can be described by a real-valued wave-equation, then wavefunctions of particles and antiparticles are equal. Such a particle (called a Majorana particle) is its own antiparticle [35]. Although the neutron is electrically neutral and has a spin of 1/2, it is not a Majorana particle because it is a binding of electrically charged quarks udd and the antineutron is $\overline{\mathrm{u}}\overline{\mathrm{d}}\overline{\mathrm{d}}$. Previously no Majorana

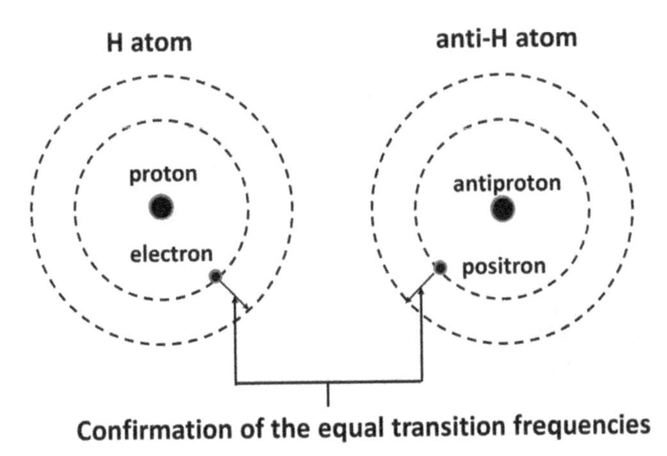

Figure 3.13. Comparison of transition frequencies of an H atom and an anti-H atom. Reproduced from [54], [55] and [56].

particles have been discovered. But there is a hypothesis that the neutrino is Majorana particle, which may be confirmed by the discovery of the neutrino-less double β-decay shown in figure 3.14 [36]; two neutrons in the same nuclear decay ($n \rightarrow p + e^{-} + \overline{\nu}_e$) simultaneously. If the neutrino is a Majorana particle, two neutrinos can be pair annihilated and it is observed as a neutrino-less double β-decay. The neutrino-less double β-decay has never been discovered, although a two neutrino double β-decay has been observed [36]. The lifetime of the neutrino-less double β-decay has been experimentally estimated to be longer than 10^{25} years [37].

The neutrino is a Fermion having the spin of 1/2, but the state of $M_S = -1/2$ (left-handed) has only been observed. It is a mystery that the state of $M_S = 1/2$ (right-handed) has never been observed. We cannot absolutely deny its existence just because it has never been observed. If it exists, there must be some reason why it has never been discovered [38]. The properties of left- and right-handed neutrinos are quite different.

The extra-ordinal small mass of the left-handed neutrino is also a mystery. It is interpreted that it is inverse proportional to the mass of the right-handed neutrino. The small mass of the left-handed neutrino may indicate that the mass of the right-handed neutrino is much larger than the proton mass.

The right-handed neutrino can decay to quarks, antiquarks, or leptons. The production rate of the particles by the decay of the right-handed neutrino might be slightly higher than that for the antiparticles. Then the unequal abundance of the particle and antiparticle is explained. The right-handed neutrino could also be a candidate for dark matter, whose identity has never been resolved (section 4.3).

3.11 Unification of forces

Section 3.3 indicates four fundamental interactions, strong, electromagnetic, weak, and gravitational. Einstein had an inspiration that these interactions are just the different sights of one interaction. He tried to unify the four interactions, but he did

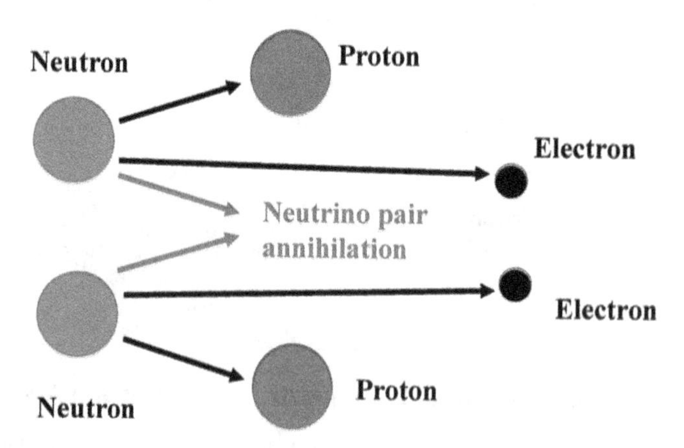

Figure 3.14. When double β-decay, two protons, electrons, and neutrinos are produced. If the neutrino is a Majorana particle, pair annihilation can be caused. The discovery of the neutrino-less double β-decay proves that the neutrino is Majorana particle.

not succeed. But his fundamental concept is still supported; he could not succeed because he started with the gravitational interaction for which the treatment is most difficult.

The electromagnetic interaction and the weak interaction were unified by Weinberg and Salam [39]. Both of these interactions have quite different properties. P-symmetry is conserved with the electromagnetic interaction, while it is violated with the weak interaction. The electromagnetic interaction is propagated by the photon whose mass is zero, while the weak interaction is propagated by the W or Z Bosons with masses much larger than protons. However, both interactions become the same ones with high energy. The non-zero mass is caused by the interaction with the Higgs particle. The mass of W and Z Bosons are zero with high energy, because Higgs particles are evaporated [40, 41]. The P-symmetry conserves with high energy, like the view from the top of a mountain as shown in figure 3.15. Losing energy, the position changes to a place with lower potential energy, just as when we descend to the foot of the mountain. What position around the mountain? The probability is uniform. But after going to a position at the foot of the mountain, the sight is no longer symmetric. After W and Z Bosons lose energy, they have non-zero mass and the P-symmetry is violated [42]. The theory of Weinberg–Salam indicates that we cannot recognize the difference between the weak interaction and the electromagnetic interaction when the energy is high enough. We call the unified interaction the 'electroweak interaction'.

A grand unified theory (GUT) is a model in which the strong interaction is also unified with the electroweak interaction at higher energy. Quarks forming protons or neutrons can be separated by ultra-high energy, and the property of the strong interaction might be the same as that for the electroweak interaction. GUT is not completed yet because the decay of the proton has not been discovered. Note that free neutrons have the β-decay ($n \rightarrow p + e^{-} + \overline{\nu}_e$) with a lifetime of 881 s. GUT predicts the possibility of decay $p \rightarrow \mu^{+} + K^{0}$ or $\overline{\nu} + K^{+}$. The mean lifetime of a proton is estimated to be 10^{31}–10^{36} years. The experiment to detect proton decay has been performed at Super-Kamiokande, having 500 million liters of distilled water at

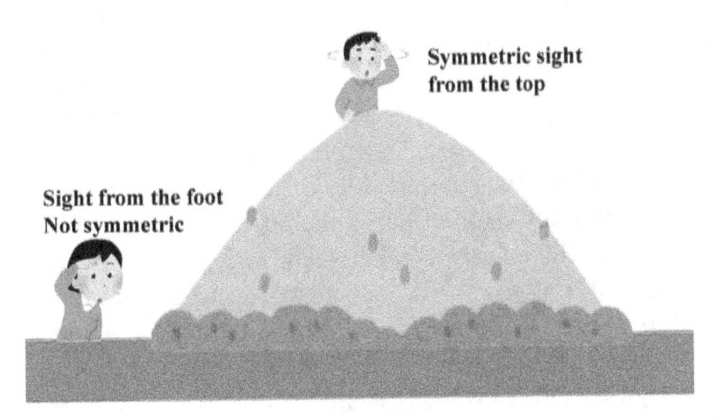

Figure 3.15. The weak interaction conserves the P-symmetry with high energy, just like the sight from the top of a mountain is symmetric. With lower energy, P-symmetry is violated, just like the sight from the foot is not symmetric.

1000 m underground in Kamioka, Gifu-prefecture in Japan. The decay of protons from one of 10^{33} water molecules might be observed per several years. The lower limit of mean lifetime was estimated to be 3.6×10^{33} years [43].

3.12 What is the gravitational interaction between particle and antiparticle?

There is a question whether antimatter is gravitationally attracted or repulsed from matter. It is also unknown whether the magnitude of the gravitational force is the same. Previously we didn't have enough experimental information.

Following the equivalence principle between the inertial force and the gravitational force, the gravitational force between antimatter and matter seems to be same as that between matter. Observing an object from an accelerating frame, we see this object accelerating in the opposite direction, regardless of whether it is matter or antimatter. The inertial force is common for matter and antimatter. If the equivalence principle is valid also for antimatter, the gravitational force should be common also for antimatter.

A recent experiment conducted at The European Organization for Nuclear Research (known by its French acronym, CERN) in Switzerland suggested that that matter and antimatter particles responded to gravity in the same way, with an accuracy of 97% [44].

The electromagnetic force to antimatter works in the inverse direction because of the conjugated electric charge. With this motion, the CPT-symmetry holds between matter and antimatter. If the gravitational force works in the same direction for matter and antimatter, the CPT-symmetry is violated while interpreting the gravitational force as an attractive force. However, interpreting the motion in the gravitational potential field as an inertial motion in a distorted coordinate, the CPT-symmetry is not violated.

3.13 The quantum electrodynamics and proton size puzzle

Here we consider the energy state of electron in a H atom. In the $n = 2$ state (n: principal quantum number), there are $(L, J) = (0, 1/2)$, $(1, 1/2)$ and $(1, 3/2)$ states, where L and J indicate the quantum numbers of electron orbital angular moment and the fine structure state. The (1, 3/2) and (1, 1/2) states indicate that the electron spin (1/2) and electron orbital angular momentum are parallel and antiparallel, respectively. Energy eigenvalues obtained by the Schrödinger equation are equal for these three states. The relativistic effect is included with the solution of the Dirac equation, which makes the energy gap between (1, 3/2) and (1, 1/2) states with the transition frequency of 10.7 GHz, but no energy gap between (0, 1/2) and (1, 1/2) states is derived as shown in equation (1.6.31). However, a slight energy gap was discovered also between (0, 1/2) and (1, 1/2) states (transition frequency 1.06 GHz) [45] as shown in figure 3.16. This energy gap is called the 'Lamb shift'. This energy gap was attributed to the interaction between the vacuum energy fluctuations and the hydrogen electron in different orbits ($L = 0$: spherical symmetric, $L = 1$: polarized). The discovery of Lamb shift played a significant role in the development

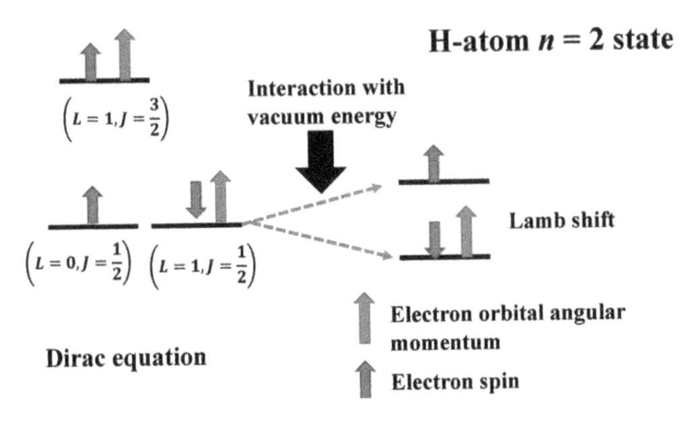

Figure 3.16. Energy gap between the ($L = 0$, $J = 1/2$) and ($L = 1$, $J = 1/2$) states of H atom in the $n = 2$ state by Lamb shift, Here, n is the principle quantum number. L and J are quantum numbers of the electron orbital angular momentum and the fine structure, respectively.

of the field of quantum electrodynamics (QED) taking the vacuum energy fluctuation into account. A correction of energy eigenvalue by the QED effect is required also in the $n = 1$ state.

QED provided an explanation for another phenomenon. The coefficient of the energy shift of an electron induced by the magnetic field (Zeeman coefficient) is 0.1% higher than the estimation from the Dirac equation. The Zeeman coefficient was measured with the uncertainty of 10^{-13} [46], which has been explained by the QED effect [47]. For the Zeeman coefficient of μ^--particle, there is discrepancy between measurement and calculation with the ratio of 2×10^{-10}, while measurement and calculation uncertainties are 4×10^{-12}.

After the establishment of the QED, the energy level of the hydrogen atom can be calculated with high accuracy. The remaining cause of the discrepancy between experiment and calculation is the finite size of the proton. It has been estimated to be $(0.8775 \pm 0.005) \times 10^{-15}$ m [48]. Its accuracy was expected to be improved by measurement with the muonic hydrogen atom (proton + μ^--particle), because the radius of orbit of μ^--particle is much smaller than that of the electron and the influence of proton size is more significant. However, the obtained result was $(0.842 \pm 0.001) \times 10^{-15}$ m [49] as shown in figure 3.17. The discrepancy between both results is larger than five times the measurement uncertainties and it is mystery of modern physics called the 'proton radius puzzle'. This puzzle might be a chance to develop new physics, including extra dimensions. Or it might be caused by the problem of the analysis of the energy structure (particularly for the QED analysis of μ^-— particle, seeing the slight discrepancy of Zeeman coefficient).

3.14 Violation of the symmetry of optical isomers of chiral molecules

The P-symmetry is violated also between the optical isomers of chiral molecules. Chiral molecules are non-superimposable on their mirror images. The mirror images of a chiral molecules are called optical isomers (figure 3.18). These molecules rotate the polarization of transmitted light, and optical isomers rotate in the inverse

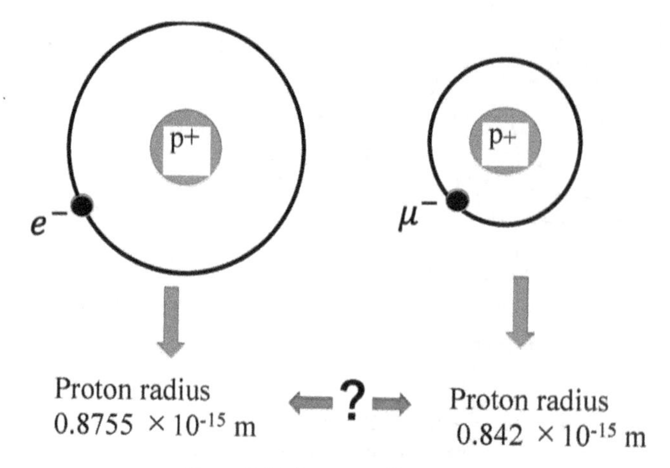

Figure 3.17. Proton radius puzzle.

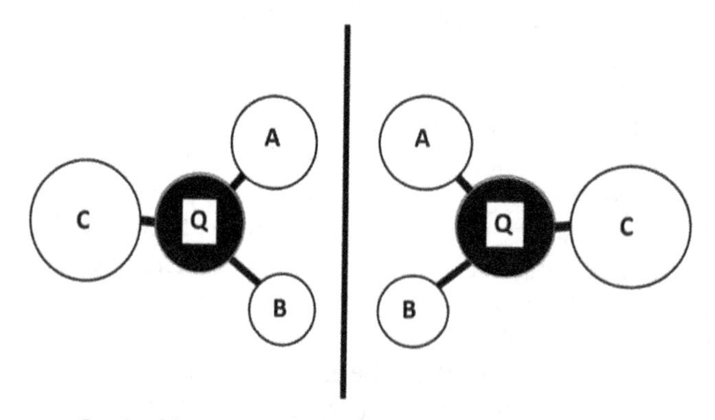

Figure 3.18. Optical isomers of chiral molecules.

direction. Therefore, individual optical isomers are designated as being either right- or left-handed. The mystery is that the abundances of right- and left-handed optical isomers are not always equal, indicating violation of the P-symmetry induced by the weak interaction [50]. The weak interaction works only within the size of the nucleus. But electrons without the orbital angular momentum can get weak nuclear force because of the non-zero distribution of the wavefunction at the position of nuclei. The violation of the P-symmetry might also be induced by the interaction with dark matter (chapter 4). If there is a slight asymmetry of the energy structure, the abundance of the optical isomer with the lower energy state is larger than that for the other.

To confirm this assumption, the slight difference in the transition frequencies must be detected. The localized abundance of optical isomers is consistent with the theoretical estimate if there is a difference in the string coefficients of atomic bonding (vibrational transition frequency) with the ratio of 10^{-14}. This accuracy is not easily attained with the transition frequencies of polyatomic molecules at room temperature. Several groups are in the process of developing experimental devices for this purpose [51]. Different methods to reduce the kinetic energy of molecules have been developed. Optoelectrical Sisyphus cooling (Appendix G) seems to be particularly useful for the cooling of chiral molecules, which makes the precision measurement of vibrational transition frequencies [52, 53].

The precise measurement of the molecular transition frequency provides a greater contribution not only for physics but also for molecular biology, because many biologically active molecules are chiral, including amino acids and sugars. Amino acids are mostly left-handed and sugar is mostly right-handed. With equal abundance of optical isomers, the appearance of organism might be something different.

If the violation of the P-symmetry is induced by the interaction with dark matter (chapter 4), the abundance ratio between optical isomers might depend on the density of dark matter. It would be useful to compare with the abundance ratio with that in other planets; for example, asteroids from where some samples have been brought back to Earth by aircraft.

References

[1] Dirac P A M A 1930 Theory of electrons and protons *Proc. R. Soc. Lond.* A **126** 360
[2] Hanson N R 1963 The Concept of the Positron (Cambridge: Cambridge University Press) p 136–9
[3] Feynman R P 1948 Space-time approach to non-relativistic quantum mechanics *Rev. Mod. Phys.* **20** 367
[4] Roentgen W C 1898 *The X-rays* Annual report of the board of Regents in Smithsonian Institution **137–99** 141–3
[5] Enz C P 1981 *Helv. Phys. Acta* **54** 412
[6] Winter K 2000 *Neutrino Physics* (Cambridge University Press) p 33
[7] Wu C S *et al* 1957 *Phys. Rev.* **105** 1413
[8] Gell-Mann M 1995 *The Quark and the Jaguar: Adventures in the Simple and the Complex* (Henry Holt and Company) p 180
[9] Lella L D and Rubbia C 2015 The discovery of W and Z particles *60 Years of CERN Experiments and Discoveries* (World Scientific)
[10] Maki Z, Nakagawa M and Sakata S 1962 Remarks on the united model of elementary particles *Prog. Theor. Phys.* **28** 870
[11] Taroni A 2015 *Nat. Phys.* **11** 891
[12] Zyla P A *Prog. Theor. Exp. Phys.* **2020** 083C01
[13] Chang K 2017 CERN *Physicists Find a Particle with a Double Dose of Charm* (The New York Times)
[14] Christenson J H *et al* 1964 *Phys. Rev. Lett.* **13** 138
[15] Angelopoulos A *et al* 1998 *Phys. Lett.* B **444** 43
[16] Schwarzschild B M 2013 *Parity* **28** 32
[17] Georgescu L 2020 *Nat. Rev. Phys.* **2** 120
[18] Kobayashsi M and Maskawa T 1973 *Prog. Theor. Phys.* **49** 652
[19] Peccei R D 2006 (ArXiv:hep-ph/0607268)

[20] Weinberg S 1978 *Phys. Rev. Lett.* **40** 223
Wilczek F 1978 *Phys. Rev. Lett.* **40** 279
[21] Hudson J J *et al* 2011 *Nature* **473** 493
[22] Baron J *et al* 2014 *Science* **343** 269
[23] Andreev V *et al* 2018 *Nature* **562** 355
[24] Roussy T *et al* 2023 *Science* **381** 46
[25] Lim J *et al* 2018 *Phys. Rev. Lett.* **120** 123201
[26] Kozyryev I *et al* 2017 *Phys. Rev. Lett.* **118** 173201
[27] Griffith W C *et al* 2009 *Phys. Rev. Lett.* **102** 101601
[28] Rosenberry M A *et al* 2001 *Phys. Rev. Lett.* **86** 22
[29] Kostelecky V A 1998 *The Status of CPT* (ArXiv:hep-ph/9810365)
[30] Hori M *et al* 2011 *Nature* **475** 484
[31] Borchert M J *et al* 2022 *Nature* **601** 53
[32] Ahmadi M 2018 *Nature* **557** 71
[33] Ahmadi M 2018 *Nature* **561** 211
[34] Baker C J *et al* 2021 *Nature* **592** 35
[35] Majorana E 1937 *Il Nuovo Cimento (in Italian)* **14** 171
[36] Rodejohann W 2011 *Int. J. Mod. Phys.* **E20** 1833
[37] Alduino C 2016 *Phys. Rev.* C **93** 045503
[38] Bayarsky A *et al* 2019 *Prog. Part. Nucl. Phys.* **104** 1–45
[39] Wineberg S 1980 *Rev. Mod. Phys.* **52** 515
[40] Higgs P W 1964 *Phys. Rev. Lett.* **13** 508
[41] Guralnik G S 1964 *Phys. Rev. Lett.* **13** 585
[42] Nambu Y and Jona-Lasinio G 1961 *Phys. Rev.* **122** 345
[43] Matsumoto R *et al* 2022 *Phys. Rev.* D **106** 072003
[44] Anderson E K *et al* 2023 *Nature* **621** 716
[45] Lamb W E and Ratherford R C 1947 *Phys. Rev.* **72** 241
[46] Odom B *et al* 2006 *Phys. Rev. Lett.* **97** 030801
[47] Brodsky S J *et al* 2004 *Nucl. Phys.* B **703** 3
[48] Kugel H W and Murnick D E 1977 *Rep. Prog. Phys.* **40** 297
[49] Sick I and Trautmann D 2014 *Phys. Rev.* C **89** 012201
[50] Quack M 2002 *Angew. Chem. Int. Ed.* **41** 4618
[51] Quack M 2008 *Ann. Rev. Phys. Chem.* **59** 741
[52] Zeppenfeld M *et al* 2009 *Phys. Rev.* A **80** 041401 (R)
[53] Prehn A *et al* 2016 *Phys. Rev. Lett.* **116** 063005
[54] Kajita M 2018 *Measuring Time* (Bristol: IOP Publishing)
[55] Kajita M 2019 *Measurement, Uncertainty and Lasers* (Bristol: IOP Publishing)
[56] Kajita M 2020 *Cold Atoms and Molecules* (Bristol: IOP Publishing)

IOP Publishing

Fundamentals of Modern Physics
Unveiling the mysteries
Masatoshi Kajita

Chapter 4

Unsolved mysteries in astrophysics and cosmology

The theory of general relativity indicates that the shape of the Universe cannot be immutable, and this conclusion was confirmed by the discovery of the expansion of the Universe. Stars are moving away with a velocity proportional to their distance from the Earth, which indicates that the Universe was born by an explosion at a point, called the 'Big Bang'. This idea was supported by the discovery of the cosmic microwave background. Still a mystery is the fact that the expansion of the Universe is accelerating driven by an unknown force called dark energy. Another mystery is that the gravitational force working in the galaxies is much larger than the estimation of the total masses of visible matter. Therefore there must be some unknown matter, called dark matter. Its identity might be clarified by observing the variation of fundamental constants; fine structure constant or proton-to-electron mass ratio. The identity of dark matter may solve the mystery of the violation of CP-symmetry. A simple introduction about the possibility of the fifth force and the superstring theory is also given in this chapter.

4.1 The shape of the Universe is not immutable

The theory of general relativity indicates that gravity is the distortion of the coordinates of space (section 1.4). In other words, the Universe cannot be described by a straight coordinate, while it contains massive objects. The size of the Universe must be reduced, and all matter should be merged in the future, as shown in figure 4.1. The shape of the Universe was believed to be immutable until the beginning of 20th century. Einstein could not accept the conclusion from his own theory and introduced a cosmological term to derivate the immutable Universe.

There was another paradox assuming the Universe with the infinite size and the immutable structure. We consider the total light intensity form all stars in the Universe with infinite size. Taking the intensity of light emitted from a star Γ_{star},

doi:10.1088/978-0-7503-6239-9ch4

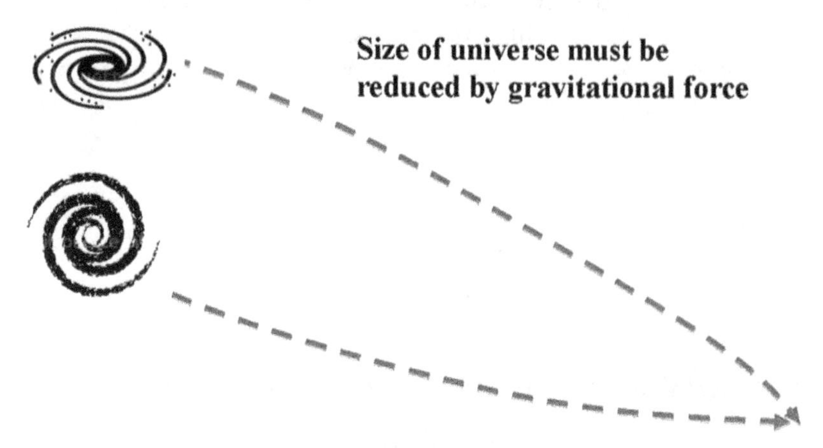

Figure 4.1. From theory of general relativity, the size of Universe must be reduced, and all matter should be merged in future.

the intensity of this light on the earth is proportional to Γ_{star}/R^2 (R: distance between the star and the earth). Assuming that the stars distribute uniformly with the density of ρ_{star} in the area of $R_{\min} < R < R_{\max}$, the total intensity is

$$I_{\text{tot}} \propto 4\pi \int_{R_{\min}}^{R_{\max}} \rho_{\text{star}} \frac{\Gamma_{\text{star}}}{R^2} R^2 dR = 4\pi\rho_{\text{star}}\Gamma_{\text{star}}(R_{\max} - R_{\min}). \tag{4.1.1}$$

Taking $R_{\max} \to \infty$, the total light intensity is infinite. It cannot be dark also at night.

This paradox was solved when Hubble's law was discovered in 1929 [1]. This law stated that the objects observed in deep space are moving away from the Earth with a velocity approximately proportional to their distance from the Earth. The motion velocities of stars are estimated by analyzing the radiation emanating from stars, which is intense or weak at some discrete frequencies. The observed frequency distribution indicates the spectrum of materials in the star. The observed spectrum is shifted by the Doppler effect, from which we can estimate the motion velocity of the star. The distances to the far distant stars are estimated from the brightness of a supernova explosion because the annual parallax is observed only for stars closer than 1500 light years. Assuming the uniform release energy of a supernova explosion, its brightness is inversely proportional to the square of the distance.

Light from distant stars is observed as infrared or microwave radiation because of the Doppler effect with velocity proportional to the distance. The total light intensity from the whole Universe is finite, and it is dark at night. Considering that the present Universe is in the process of expanding, there is no discrepancy with the theory of general relativity. Einstein canceled the introduction of the cosmological term.

With the time reversal of the motion with the relative velocities proportional to the distance, the Universe converges to one point at 13.8 billion years ago, as shown in figure 4.2. The Universe was born by the explosion at one point, called the Big Bang. This idea was supported by the discovery of the cosmic microwave background (CMB) in 1965 [2]. The frequency distribution of the CMB corresponds to

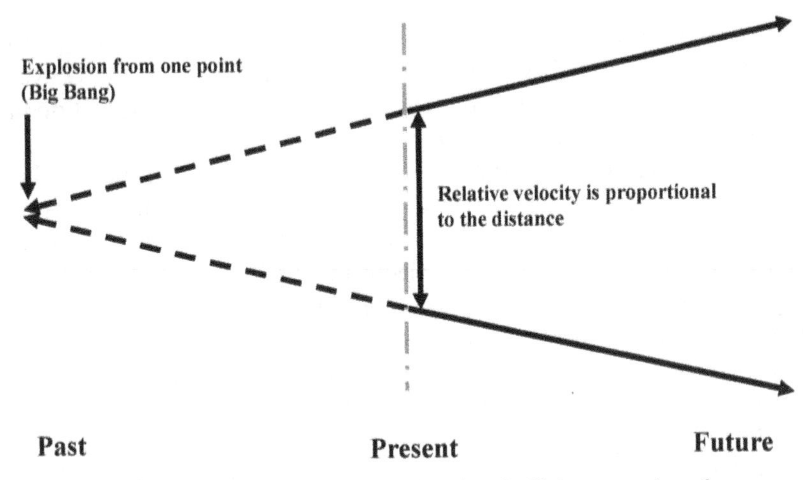

Figure 4.2. Simple description of Hubble's law, indicating that the Universe was born from an explosion at one point, called the Big Bang.

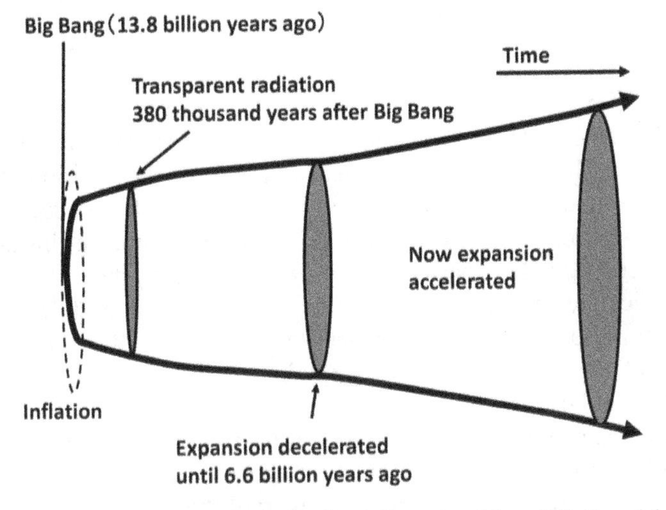

Figure 4.3. The history of the Universe since the Big Bang. Reproduced from [52]. Copyright IOP Publishing Ltd. All rights reserved.

the blackbody radiation with an absolute temperature of 2.725 K in all directions within an accuracy of 10^{-5}. The uniform temperature shows that the Universe was a single object in the past. The temperature of the Universe during the Big Bang was unimaginably high. Soon after the Big Bang (10^{-36}–10^{-34} s), there was a phase transition from a high energy vacuum to a low energy vacuum and an expansion was induced with a speed much faster than the speed of light (called inflation) [3, 4], as shown in figure 4.3. The theory of relativity prohibits motion in a space faster than the speed of light, but it does not prohibit the expansion of the space itself faster than the speed of light. As the Universe expanded, the temperature decreased. After 380 000 years, the temperature was around 3000 K, commencing a period when atoms were formed from nuclei and electrons. After this period, the Universe became transparent to electromagnetic waves (the transparent radiation of the Universe) and the CMB is understood to be the radiation from this period. The CMB observed on Earth is redshifted by the continued expansion of the Universe.

Observation of quasars at long distance is useful for research into the history of the Universe. Observing a quasar at a distance of 13 billion light years, we can get the information at 13 billion years ago. Before the transparent radiation of the Universe cannot be observed with the electromagnetic wave but can be observed with the gravitational wave. The signal shape of the gravitational wave at the birth of the Universe is expected to be quite different from the waveform observed with the merging of black holes or neutron stars. For example, its wavelength is expected to be much longer than those observed previously (section 2.10) because of the Doppler effect by the expansion of the Universe (same reason that CMB is observed with the microwave region). Therefore, it is very difficult to detect with the laser interferometry system on the Earth (cavity length of several km). There is a project to construct a laser interferometry system using cavities with lengths of 20 000 km constructed by mirrors in satellites.

4.2 Dark energy

It is reasonable to expect that the velocity of expansion decreases by the gravitational interaction. However, precision measurements of the positions and velocities of galaxies show that the expansion of the Universe has been accelerating for the past 6.62 billion years as shown in figure 4.3. The force causing this acceleration is called 'dark energy' [5]. The identity of dark energy is a mystery of modern physics. The cosmological term, which Einstein once introduced and later canceled, is introduced again to describe the present expansion of the Universe. Detailed measurement of the gravitational wave might provide useful information to find the identity of the dark energy.

The precise measurement of the temperature distribution of the CMB may provide useful information concerning dark energy. The temperature distribution of the CMB was believed to be homogeneous until the 1980s, but the fluctuation at levels of 10^{-6} was clarified after the improvement of measurement accuracy [6]. The inhomogeneous polarization of CMB was also discovered. This fluctuation derives from the fluctuation in the density of matter in the early Universe, and it was imprinted shortly after the Big Bang. This result gives much information about the origin of galaxies and large structure of the Universe. The detailed information of non-homogeneity of CMB also gives information about the inflation soon after the Big Bang, which provides a chance to solve the mystery of dark energy (we don't know if the inflation was caused by the same force as dark energy).

Many researchers have been eager to search for the identity of dark energy. On the other hand, a new hypothesis was given to deny the existence of dark energy. Lee *et al* insists that the estimation of the distance to the stars from the brightness of the supernova explosion is questionable [7]. The release energy by supernova explosion might depend on the time or position because there might be a change of circumstance. With their corrected estimations of distances, no acceleration was indicated. However, more investigation will be required into the reliability of the corrected distance.

4.3 Dark matter

4.3.1 Mystery of dark matter

The phrase 'dark matter' was used in the papers by Kapteyn [8] and Oort [9] to describe the motion of stars close to the solar system. In 1933, Zwicky assumed the existence of dark matter to explain the missing mass of the galaxy to describe the velocity of galaxies in a galaxy cluster [10]. From the balance between the centrifugal force and the gravitational force,

$$\frac{GM_{\mathrm{Gal}}^2}{R_{\mathrm{Gal}}^2} = \frac{M_{\mathrm{Gal}}V_{\mathrm{Gal}}^2}{R_{\mathrm{Gal}}}$$
$$M_{\mathrm{Gal}} = \frac{R_{\mathrm{Gal}}V_{\mathrm{Gal}}^2}{G}, \tag{4.3.1}$$

is derived. Here, R_{Gal} is the distance between galaxies, M_{Gal} is the mass of galaxy, V_{Gal} is the relative velocity between galaxies, and G is the gravitational coefficient. Zwicky observed the relative velocity between galaxies in the Coma cluster to be $1000 \mathrm{~km~s^{-1}}$. The size of the Coma cluster was estimated to be 3.3 million light years. The mass of clusters was estimated to be 3×10^{14} M_{sol} (M_{sol}: solar mass), which is two orders larger than the total mass of visible stars. The same result was obtained by observation of other galaxy clusters. However, Zwicky mentioned not only the existence of dark energy but also the possibility that the Newtonian gravitational law might not be valid at large distances.

The existence of dark matter is determined from the velocity of H atoms at the edges of galaxies [11]. The measured velocity is much higher than expected and the gravitational force required to balance the centrifugal force is ten times larger than that given by the visible matter within each galaxy, as shown in figure 4.4. The nature of this unknown substance remains a mystery.

Verlinde published a paper which denied the existence of dark matter. In this paper, he insists that the phenomena shown above are caused by the interaction between conventional matter and dark energy [12]. Kroupa also denies the existence of dark matter insisting that the deceleration of motion of large matter has never been observed which must be caused while moving in a cloud of dark matter [13]. They argue that the mysteries of the gravitational interaction are solved by modification of the theory of the gravitational force. However, these theories have discrepancies with the structures of galaxies, which were discovered afterward. The main proof of the existence of dark matter is that the mass distribution measured by the gravitational lens effect (section 1.4) are not consistent with the distribution of visible matter [14].

In the 1980s, the distribution of groups of galaxies was mapped and a bubble-like structure of the Universe appeared (figure 4.5) [15], along with areas called voids,

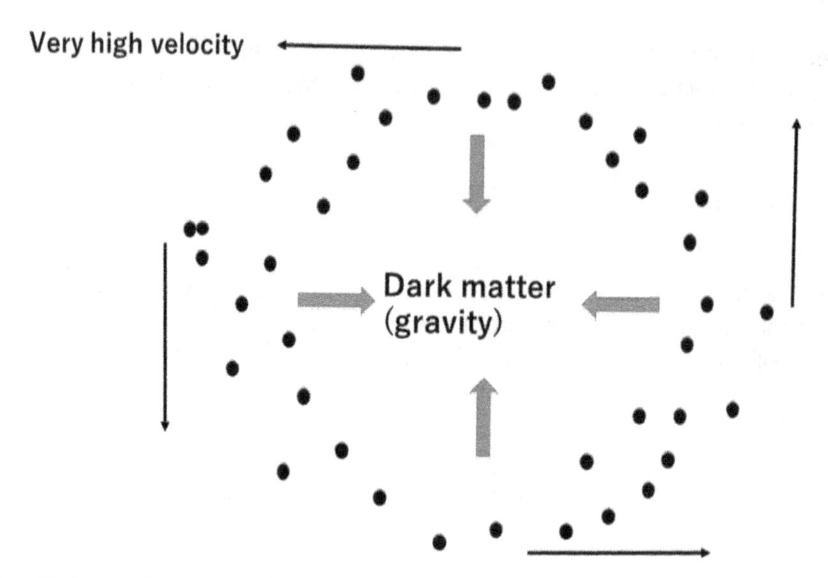

Figure 4.4. H atoms at the end of galaxies are moving very fast and the gravitational force from dark matter is required to balance the centrifugal force. Reproduced from [52].

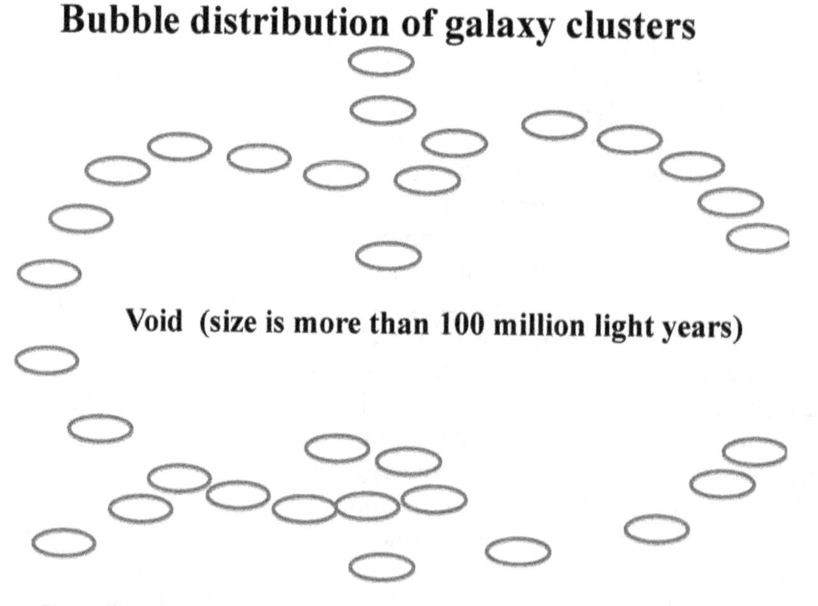

Figure 4.5. Distribution of galaxy clusters appear to have a bubble-like structure.

where there are no galaxies. The sizes of these voids are larger than 100 million light years. This structure shows that the mass of the Universe is dominated more by dark matter than visible matter. It was believed that no galaxies or stars can be constructed without dark matter. Recently, a galaxy without dark matter

NGC1052-DF_2 was observed [16], which created a new mystery for the construction of galaxies.

4.3.2 What is the identity of dark matter?

Several candidates have been considered as the identity of dark matter. The neutrino was considered as one candidate since its non-zero mass has been identified, because it has no electromagnetic interaction. This hypothesis was given assuming that neutrino mass is larger than $2 \times 10^{-5} m_e$ (we don't know the mass). This theory is not supported recently, because its distribution cannot be localized in an area smaller than a galaxy cluster. To give a large gravitational force at the center of a galaxy, dark matter should be localized at a limited area in a galaxy. However, the right-handed neutrino (section 3.10) might be a candidate if it exists (it has not yet been discovered).

The weak interactive massive particle (WIMP) is another candidate [17]. This is a lightest supersymmetric particle with neutral electric charge. The mass is estimated to ne 1–100 m_p. This particle has been theoretically predicted to have a weak interaction, because the gravitational interaction is too weak for the thermalization of the early Universe. Experimental efforts to detect WIMPs have been performed including the search for expected products at the annihilation of WIMP (gamma rays, neutrinos etc) in nearby galaxies and galaxy clusters. Direct detection experiments were also designed to measure the collision of WIMPs with nuclei in the laboratory, as well as attempts to directly produce WIMPs in colliders. However, this particle has not been discovered and it makes the simplest WIMP hypothesis doubtful. The weak interaction with a WIMP might cause the violation of the P-symmetry of atoms or molecules.

An axion with zero spin and mass of $10^{-11} - 10^{-9} m_e$ is now the most advantageous candidate, although it has not yet been discovered. This particle was proposed to solve the strong CP-problem (section 3.7). The de Broglie wavelength has the same order as the size of galaxy, therefore axions are expected to be in the Bose–Einstein-condensation state (section 2.9.2) also at room temperature. The experimental efforts to detect the axion are described in sections 4.4 and 4.5.

Except for unknown particles, some heavenly bodies can be candidates for dark matter. For example, black holes cannot be observed with light and could be a candidate, although their total mass is estimated to be lower than 40% of the total mass of dark matter. Neutron stars with weak light could also be candidates. Planets, which have not been observed, might be candidates.

The rotation of the polarization of CMB induced by the interaction with dark matter has been observed with a reliability of 99.2% [18], which indicates the violation of the CP-symmetry. There could be more than one answer about the identity of dark matter, but it is expected to include some particle which induces the violation of the CP-symmetry. More detailed research of CMB might give us the information required to clarify the properties of dark matter.

4.4 Search of variation in fundamental constants

4.4.1 Is there no spatial or temporal variation with fundamental constants?

Physical laws are established based on many fundamental constants. With different values of the fundamental constants, the appearance of the Universe would be quite different. If the ratio of the electromagnetic force to the strong nuclear force is higher, atoms with heavy nuclear could not exist because of the repulsive force between protons. With smaller electromagnetic force, molecular bonding is not possible. The relativistic effects are more significant with lower speed of light and the quantum effects are more significant with a larger value of Planck constant. We would not exist with any other combination of fundamental constants. The combination of suitable fundamental constants looks like too much of a coincidence. If fundamental constants have a dependence on time and position, we may understand that we are living in an epoch with suitable combinations of fundamental constants. In 1937, Dirac mentioned for the first time the possibility (not requirement) of time-varying fundamental constants [19]. However, its effect was expected to be too small to detect with the measurement uncertainty at that time. Currently, some transition frequencies can be measured with uncertainties lower than 10^{-16} and the search for such variations has become an active topic of investigation. The variation in fundamental constants might indicate some information of the dark energy or dark matter.

When there is a variation in a fundamental constant X, the energy structure of atoms and molecules change. The transition frequencies change with different ratios, as shown in figure 4.6. The variation in X (ΔX) is determined from the variation in the ratio of two transition frequencies $\nu_1(\propto X^{K_{X1}})$ and $\nu_2(\propto X^{K_{X2}})$,

$$\frac{\Delta(\nu_1/\nu_2)}{(\nu_1/\nu_2)} = (K_{X1} - K_{X2})\frac{\Delta X}{X}. \tag{4.4.1}$$

A non-zero $(\Delta X/X)$ is detected when $\Delta(\nu_1/\nu_2) > \delta(\nu_1/\nu_2)$, where $\delta(\nu_1/\nu_2)$ is the measurement uncertainty in (ν_1/ν_2). The minimum detectable value of $(\Delta X/X)$ is

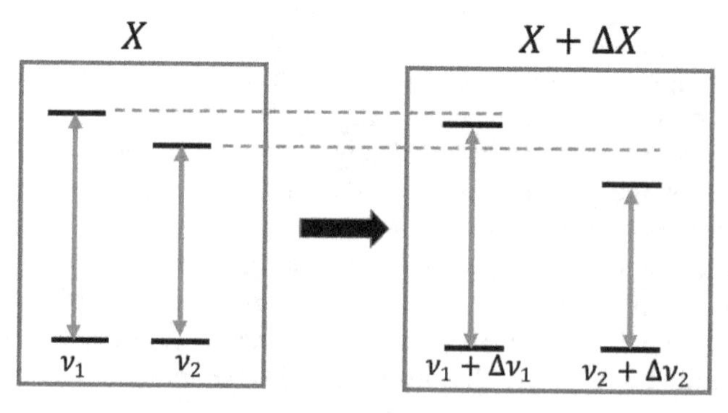

Figure 4.6. When a fundamental constant changes, atomic or molecular transition frequencies change with different ratios.

$$\left(\frac{\Delta X}{X}\right)_{\min} = \frac{1}{K_{X1} - K_{X2}} \frac{\delta(\nu_1/\nu_2)}{(\nu_1/\nu_2)} \tag{4.4.2}$$

In the selection of the transition frequencies to monitor the variation in X, we have a choice between a high value in the sensitivity parameter K_X or an attainable measurement accuracy in the transition frequency. Two methods are possible in the search for the variation in the fundamental constants. One is a comparison of (ν_1/ν_2) for a quasar and for the Earth to see the variations over a long period. The values in a quasar at a distance of 10 billion light years are those at 10 billion years ago. With observations from quasars in different directions, we can distinguish between spatial and temporal variations. The problem with this method is that the measurement accuracies of transition frequencies in quasars are limited. The other method is the measurements of (ν_1/ν_2) in a laboratory with the uncertainty below 10^{-16}. We cannot distinguish between a spatial variation and a temporal one only with the measurement in a laboratory, because the solar system moves in the Universe with a velocity of 240 km s^{-1} (one order faster than the Earth's orbital velocity). On the other hand, the quick variation (e.g., induced by interaction with dark matter) of fundamental constants can be detected by the measurement in a laboratory. In this section, we consider the variation in the fine structure constant $\alpha = e^2/2\varepsilon_0 hc (= 0.007\,297)$ and the proton-to-electron mass ratio μ_{pe} $(= m_p/m_e)$. If there is a variation in the fine structure constant α, there should also be a variation in the proton-to-electron mass ratio μ_{pe}, because the proton mass is dominated mainly by the binding energy (two orders larger than the total mass of quarks as shown in section 3.3). The slight variation in the electromagnetic force between quarks ($\propto\alpha$) induces the change of the binding energy, which is observed as the change of the proton mass. Reference [20] indicates the relation,

$$\frac{\Delta\mu_{pe}}{\mu_{pe}} = R_{var}\frac{\Delta\alpha}{\alpha} \tag{4.4.3}$$

where R_{var} is a constant between 28 and 40, given by the detail of grand unification theory (GUT). Comparing the variations in α and μ_{pe}, we can obtain useful information for GUT.

4.4.2 Sensitivity on the variation in fundamental constants

While the relativistic effects with the electron motion in atoms are negligibly small, the transition frequencies between different electron energy states are proportional to $m_e\alpha^2$. The ratio between transition frequencies without relativistic effects is constant also when there is a variation in α. Taking the relativistic effect into account, the relation between energy eigenvalues and α is given by

$$E_e \propto \sum_{n=1}^{\infty} c_n \alpha^{2n} \tag{4.4.4}$$

and the effect of $c_n(n \geqslant 2)$ is significant for atoms in a heavy nucleus as shown in equations (1.6.31) and (1.6.32). The variation in α is detected from the variation in the ratio between transition frequencies of heavy and light atoms. The transition frequencies of highly charged ions [21] and nuclear transition frequencies [22] are expected to be very sensitive to the variation in α. These transition frequencies are advantageous also for the precision measurement because of the small sensitivities to the perturbation given by the fluctuation of electromagnetic fields (Appendix C). However, measurements of these transition frequencies are technically complicated. It may take a longer time to realize the precision measurements of these transition frequencies.

It is difficult to search for the variation in μ_{pe} from the comparison between the atomic transition frequencies in the optical region, because of $K_{\mu_{pe}} < 10^{-4}$ (reduced mass between electron and nucleus is almost equal to m_e). Molecular vibrational rotational transition frequencies, given by the nuclear motion, have sensitivities to variations in μ_{pe}. As shown in figure 4.7, the vibrational energy is given by (Appendix D)

$$E_{\text{vib}}(n_{\text{vib}}) = \left(n_{\text{vib}} + \frac{1}{2}\right)h\nu_{\text{vib}}$$

n_{vib}: vibrational quantum number

ν_{vib}: vibrational motion frequency (4.4.5)

and the $n_{\text{vib}} \rightarrow n_{\text{vib}} + 1$ vibrational transition frequency is roughly given by $[E_{\text{vib}}(n_{\text{vib}} + 1) - E_{\text{vib}}(n_{\text{vib}})]/h = \nu_{\text{vib}}$. The rotational energy is given by

$$E_{\text{rot}}(J) = \frac{h^2 J(J+1)}{2(2\pi)^2 I_{\text{mol}}}$$

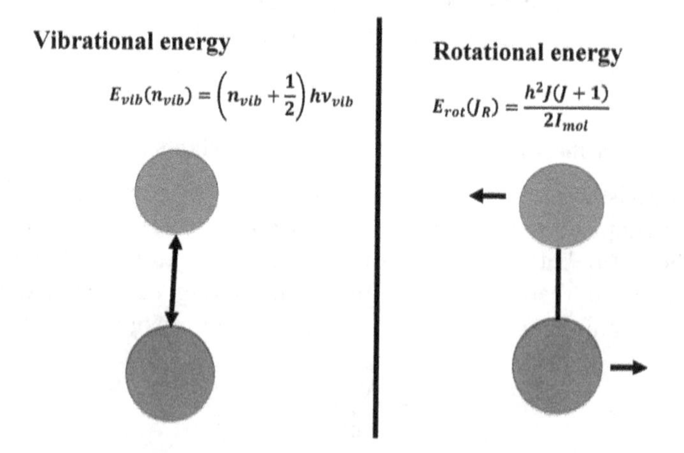

Figure 4.7. Molecular energy by vibrational or rotational motion.

J: molecular rotational quantum number

$$I_{\text{mol}}\text{: molecular moment of inertia} \tag{4.4.6}$$

and the $J \to J+1$ rotational transition frequency is given by $\nu_{\text{rot}} = [E_{\text{rot}}(J+1) - E_{\text{rot}}(h)]/h = \frac{h(J+1)}{(2\pi)^2 I_{\text{mol}}}$.

Here, we consider the dependence of the atomic and molecular transition frequencies on m_{p} and m_{e}. The Bohr radius a_B is proportional to m_{e}^{-1} and the frequency standard is given by an atomic transition frequency standard,

$$\nu_{\text{e}} \propto e^2/a_B \propto m_{\text{e}}. \tag{4.4.7}$$

The bonding length between atoms is proportional to the Bohr radius a_B, The vibrational transition frequency is given by the change in the electronic energy with the change in interatomic distance. Therefore,

$$\frac{1}{a_B} \propto m_{\text{p}} \nu_{\text{vib}}{}^2 a_B{}^2$$

$$\nu_{\text{vib}}{}^2 \propto \frac{1}{m_{\text{p}} a_B{}^3} \propto \frac{m_{\text{e}}{}^3}{m_{\text{p}}}$$

$$\frac{\nu_{\text{vib}}}{\nu_e} \propto \mu_{\text{pe}}^{-1/2}. \tag{4.4.8}$$

The moment of inertia is proportional to $m_p a_B^2$ and the rotational transition frequency is proportional to $\left(m_p a_B^2\right)^{-1}$, as shown above. Therefore,

$$\frac{\nu_{\text{rot}}}{\nu_{\text{e}}} \propto \mu_{\text{pe}}^{-1} \tag{4.4.9}$$

is derived. Molecular transition frequencies between splitting states given by the tunneling effect of atoms have higher sensitivities to the variation in μ_{pe}. For example, $K_{\mu_{\text{pe}}} = -4.3$ with ammonia inversion transition frequency [23].

4.4.3 Astronomical search for the variation in fundamental constants

The astronomical search for the variation in α has been performed comparing the spectra of several kinds of metal ions (Si^+, Fe^+, Al^+ etc) in quasars and the laboratory sample spectra. The first paper to report that the fine structure constant α in quasars (about 10 billion light years distant) is shifted by $\Delta\alpha/\alpha = (-1.1 \pm 0.4) \times 10^{-5}$ was by Webb [24]. Murphy reported the observation of $\Delta\alpha/\alpha = (-0.57 \pm 0.11) \times 10^{-5}$ [25]. Since these reports were published, the astronomical search for the variation in α became a hot subject for researchers. However, recent results after improving measurement accuracies indicate almost null results. The variation in the fine structure constant at 13 billion years ago was obtained to be $\Delta\alpha/\alpha = (-2.18 \pm 7.27) \times 10^{-5}$ by observing the spectra of Si^+, Al^+, Fe^+, and Mg^+ ions in the quasar J1120 + 0641 [26]. Observing the Fe^+ spectrum in the quasar J110325-264515 (10 billion light years distant), $\Delta\alpha/\alpha = (1.56 \pm 1.78) \times 10^{-6}$ was obtained [27].

The astronomical search for the variation in μ_{pe} was also performed. From the H_2 molecular transition frequencies in the quasar source J2123-005 (12 billion light years distant) observed by two different telescopes, $\Delta\mu_{pe}/\mu_{pe} = (8.3 \pm 4.2) \times 10^{-6}$ and $(5.6 \pm 6.2) \times 10^{-6}$ were obtained [28]. A value of $\Delta\mu_{pe}/\mu_{pe} = (0 \pm 1) \times 10^{-7}$ was obtained from the measurement of the CH_3OH transition frequency in a quasar PKS1830-211 (8 billion light years distant) [29].

Previous astronomical results indicate null results about the linear drift of the fundamental constants over these 13 billion years. However, we don't know about the physics soon after the Big Bang.

4.4.4 Search for the variation in fundamental constants by laboratory measurement

Since the atomic transition frequencies were measured with an uncertainty below 10^{-16}, the search for the variation in α has been performed comparing the following transition frequencies.

Comparison between

(1)

$^{199}Hg^+$ $(S = 1/2,\ L = 0,\ J = 1/2) \rightarrow (S = 1/2,\ L = 2,\ J = 5/2)$ $K_\alpha = -3.2$

$^{27}Al^+$ $(S = 0,\ L = 0,\ J = 0) \rightarrow (S = 1,\ L = 1,\ J = 0)$ $K_\alpha = 0.008$

$(d\alpha/dt)/\alpha = (-1.6 \pm 2.3) \times 10^{-17}$ (per year) [30],

(2)

$^{171}Yb^+$ $(S = 1/2,\ L = 0,\ J = 1/2) \rightarrow (S = 1/2,\ L = 3,\ J = 7/2)$ $K_\alpha = -6.0$

$^{171}Yb^+$ $(S = 1/2,\ L = 0,\ J = 1/2) \rightarrow (S = 1/2,\ L = 2,\ J = 3/2)$ $K_\alpha = 1.0$

$(d\alpha,/, dt)/\alpha = (-2.0 \pm 2.0) \times 10^{-17}$ (per year) [31]

$(d\alpha,/, dt)/\alpha = (-0.5 \pm 1.6) \times 10^{-17}$ (per year) [32],

$(d\alpha,/, dt)/\alpha = (1.0 \pm 1.1) \times 10^{-18}$ (per year) [33],

where S, L, and J are the quantum numbers of electron spin, electron orbital angular momentum, and the fine structure induced by equation (1.6.27).

The variation rate in α is below 10^{-18} per year. When the transition frequencies of highly charged ions [21] or nuclear transition frequencies [22] are measured with the uncertainty of 10^{-18}, a fractional variation in α of 10^{-19} can be detectable. We may determine the non-zero variation or maybe just reduce the upper limit.

The precision measurement of molecular vibrational rotational transition frequencies is useful to search for the variation in μ_{pe}. However, the molecular transition frequencies have never been measured with uncertainty below 10^{-15}. Currently $(d\mu_{pe}/dt)\mu_{pe}$ has been estimated from a comparison between the Cs hyperfine transition frequency (introduced in sections 2.1 and 2.8, which is proportional to $\alpha^{2.83}/\mu_{pe}$ and the atomic transition frequencies in the optical region. Results of $(-0.5 \pm 1.6) \times 10^{-16}$per year [31], $(0.2 \pm 1.1) \times 10^{-16}$ per year [32] and

$(-0.8 \pm 3.6) \times 10^{-17}$ per year [33] were obtained by groups in Germany and the UK, respectively. It is not realistic to reduce the measurement uncertainty of the Cs hyperfine transition frequency to below 10^{-16}. Therefore, detection of the variation in μ_{pe} with this method is difficult. The uncertainty of 10^{-18} is theoretically estimated to be attainable with N_2^+ and O_2^+ vibrational transition frequencies [34–36]. After the realization of precision measurement of molecular transitions, variation in μ_{pe} might be detected.

4.4.5 Search for dark matter from the variation in fundamental constants

Although null results of the linear drift of fundamental constants have been given, there are possibilities to search for the vibrational change of the fundamental constants from the interaction with dark matter. Here we assume the axion as the identity of dark matter. The axion is expected to be a particle with zero spin and mass m_ϕ below 10^{-6} m_e, then the axion is in the Bose–Einstein-condensation also with room temperature. The rest energy is given by $m_\phi c^2$ (section 1.3) and the wavefunction with a uniform phase (by Bose–Einstein condensation shown in section 2.9) is given by

$$\Phi \propto \cos\left[2\pi \nu_{DM} t\right]$$

$$\nu_{DM} = \frac{m_\phi c^2}{h} \qquad (4.4.10)$$

The interaction with dark matter can induce the oscillational variation in the fundamental constants, whose frequency indicates the mass of the dark matter as shown in figure 4.8. The change of the density of dark matter is observed by the change of the amplitude.

A certain measurement time τ_{mes} is required to reduce the measurement statistic uncertainty (Appendix C). Therefore,

$$\nu_{DM} \tau_{mes} \ll 1$$

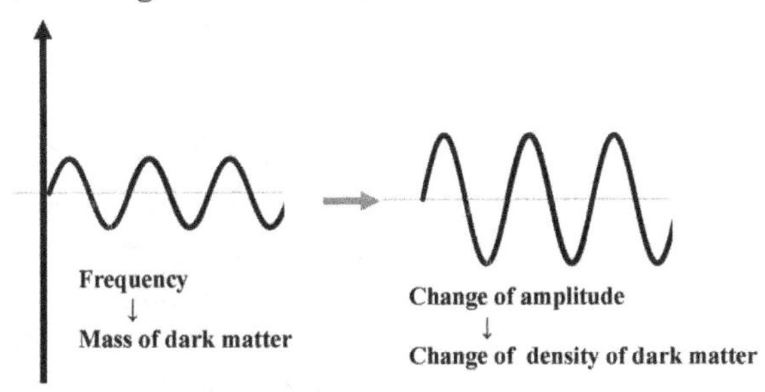

Figure 4.8. The vibrational variation of fundamental constants induced by dark matter. The frequency indicates the mass of the dark matter particle. The change of the amplitude indicates the change of density of dark matter.

Table 4.1. List of transition frequencies to be linked aiming the search for the variation of fine structure constant α and the proton-to-electron mass ratio μ_{pe}. The sensitivities for the variations in both parameters are also listed [40].

Transition frequencies	K_α	$K_{\mu_{pe}}$
^{133}Cs hyperfine transition	2.83	−1
^{87}Sr $(S = 0, L = 0, J = 0) \rightarrow (S = 1, L = 1, J = 0)$	0.06	0
$^{171}Yb^{+}$ $\left(S = 1/2,\ L = 0, J = \frac{1}{2}\right) \rightarrow \left(S = 1/2,\ L = 3, J = \frac{7}{2}\right)$	−6.0	0
$^{14}N_2^{+}$ or $^{15}N_2^{+}$ vibrational transition	0	−0.5
$^{40}Ca^{19}F$ vibrational transition	0	−0.5
Cf^{15+} $^{2}F_{5/2} \rightarrow {}^{2}I_{9/2}$,	47	0
Cf^{17+} $5f_{5/2} \rightarrow 6\ p_{1/2}$,	−45	0

$$m_\phi \ll \frac{h}{c^2 \tau_{\text{mes}}} \tag{4.4.11}$$

is required to observe the oscillational variation of fundamental constants. The experiments to observe the oscillational variations in fundamental constants is in process in several groups [36–39], but the oscillational variation has not yet been observed. The upper limit of their amplitude was determined with a wide range of ν_{DM}. With a lower frequency area, the determined upper limit is lower, because the statistical uncertainty is reduced by taking a longer measurement time. In the UK, there is a project called QSNET to establish a link to search for the variations in α and μ_{pe} by comparing the transition frequencies listed in table 4.1 [40].

The electron mass m_e is expected to be most sensitive to the interaction with dark matter, and the measurement of vibrational transition frequencies of N_2^{+} molecular ion or CaF molecule with the uncertainty of 10^{-18} might make a great contribution to the search for the identification of dark matter.

4.5 Direct detection of axion

There is also a project called XENONnT to detect the axion from the collisional interaction with liquid xenon (Xe) [41]. There might be a small but non-zero possibility that axions collide with Xe atoms in a liquid state as shown in figure 4.9(a). Then Xe atoms may be excited, which can be detected by fluorescence or electronic rebound, just like the neutrino was detected by its interaction with water molecules. There is a problem that the electronic rebound can be caused not only by the collision with axions, but also by the radioactive impurity. In 2017, electronic rebound higher than the estimated background was detected using liquid Xe of 3.2 t, and it was hoped that this result might be the discovery of the axion. However, the observed signal was concluded to be the background after the reduction of the background using a new apparatus with liquid Xe of 8.3 t [42]. Despite further efforts to reduce the background, no signal induced by the collision

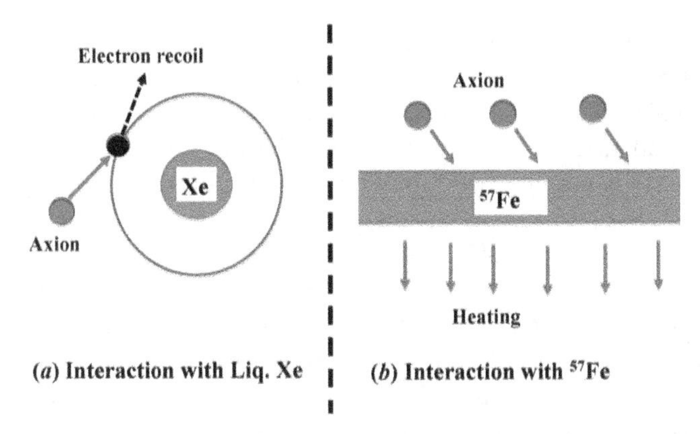

Figure 4.9. Panel(a) denotes the method to detect axions from the interaction with liquid Xe. When there is an interaction, an electron recoil is expected [39–41]. Panel (b) denotes the method to detect axions by the heating effect when axions interact with ^{57}Fe.

with axions has been observed. The upper limit of the cross section of the collisional interaction was estimated to be 2.58×10^{-47} cm^2 [43].

There is also a project to detect axions from the interaction with ^{57}Fe atoms as shown in figure 4.9(b). Axions are expected to be produced in the Sun via the interaction between photon and ^{57}Fe atoms [44]. If this theory is right, axions can interact with ^{57}Fe atoms again and give a thermal energy of 14.4 keV, which can be detected with a calorimeter.

There have been also trials to observe the conversion of astrophysical axions to microwaves using a strong magnetic field [45]. The mass of the axion is determined from the frequency of the microwave. The axion has never been discovered with this method.

As shown above, many researchers are eager to discover the axion, because it is an influential candidate for dark matter. Although it is questionable to assume the axion is the main identity of dark matter, it seems more reasonable than Fermion particles, because Fermion particles cannot have high density to give a large gravitational effect. It seems reasonable to assume the Bose–Einstein-condensation (superfluidity), so that it does not give a friction force to decelerate the motion of stars (Kroupa denied the existence of dark matter for this reason [13]). The mass is not known, although we can roughly estimate the de Broglie wavelength from the area of distribution (measured by the gravitational lens effect). The discovery of the axion may solve the mystery of the 'strong CP-problem' as shown in section 3.7. The non-zero amplitude of the oscillation of the vacuum angle might derive the violation of the CP-symmetry larger than the estimation by Kobayashi–Maskawa theory (section 3.6).

4.6 Does the fifth force exist?

Current physics is established on four kinds of interaction. However, some speculative theories have proposed a fifth force to explain various anomalous observations that do not fit existing theories. The characteristics of this fifth force depend on the hypothesis being advanced. Many postulate a force roughly the strength of gravity with a range of anywhere from less than a millimeter to cosmological scales. The search for a fifth force has increased since the discoveries of dark energy and dark matter. The fifth force between matter with the mass of $m_{i,j}$ is expected to have a potential energy with the formula of Yukawa potential,

$$V_{ij}(r) = G\frac{m_i m_j}{r}\xi_{ij}e^{-\frac{r}{\lambda}} \tag{4.6.1}$$

and currently $|\xi_{ij}| < 10^{-4}$ has been confirmed. Therefore, the fifth force is expected to be very weak and difficult to test experimentally. The first report about the discovery of the fifth force (range 100 m) was reported by Fischbach *et al* [46]. But Fischbach himself published a paper mentioning that there is no evidence of the fifth force [47].

Cartlidge reported the possibility of the fifth force from an anomalous radioactive decay of ^{8}Be atom [48]. They insist that ^{8}Be atom in an excited state is deexcited emitting an unknown Boson particle with the mass of 30 m_e and this particle decays to a pair of an electron and a positron. Feng *et al* gave an interpretation that this anomaly's radioactive decay is caused by the fifth force [49]. This force can be unified with the electromagnetic force, strong interaction, and weak interaction. We need more information to conclude whether it is really the fifth force.

4.7 Does the extra dimension exist?

The unification of the theory of general relativity and quantum mechanics has never been attained. With the limit of zero-distance, the energy becomes infinite because of the uncertainty principle between position and momentum. Superstring theory has been proposed to resolve these questions [50]. This theory considers particles to be a string having a finite size (figure 4.10) rather than being point-like. Having finite size,

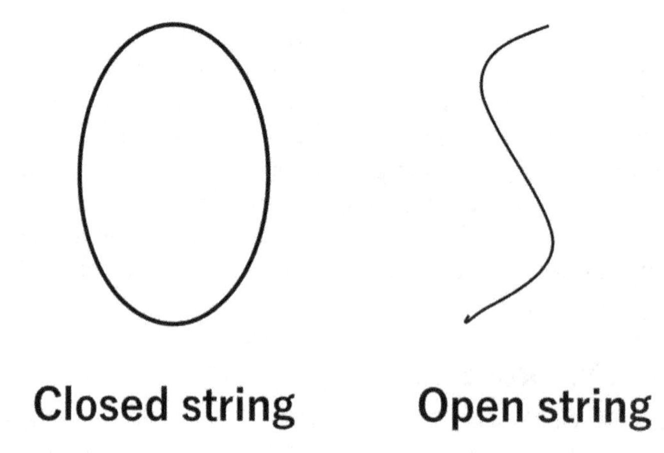

Figure 4.10. Schematic of particles as strings of finite length in the superstring model. Reproduced from [52].

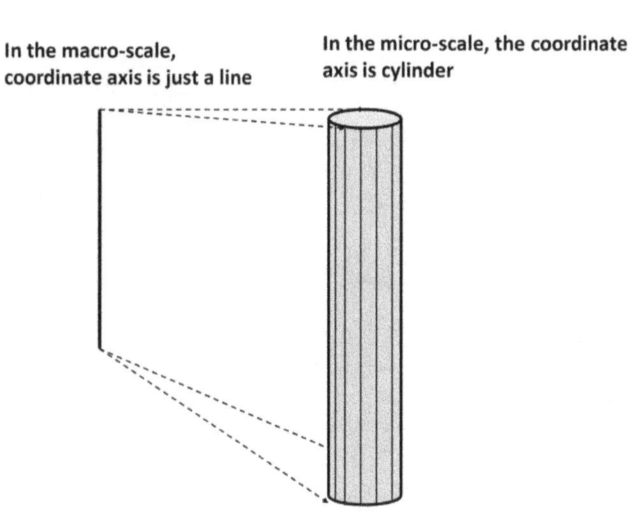

Figure 4.11. Schematic of extra dimensions. A line viewed at a distance looks one-dimensional but may be cylindrical with a finite radius when viewed close-up. Reproduced from [52]. Copyright IOP Publishing Ltd. All rights reserved.

the energy is not infinite. An open string includes photon, W particle, Z particle, and gluons with spin 1. A closed string includes graviton with spin 2. Particle mass energy is given by the vibrational energy of the string. It is very difficult to give experimental confirmation because the size of the string is expected to be as small as the order of 10^{-35} m, called the 'Planck length'.

The superstring theory requires ten or more dimensions, although we know only four dimensions (positions in three directions and time). Then, where are the extra dimensions? We cannot give a clear explanation. The Big Bang might have induced expansion only in the four dimensions, while the size in extra dimensions is small. The coordinate axis are lines at macro-scale, but they may be cylinders having an extra dimension at micro-scales (figure 4.11). The gravitational force can interact with extra dimensions, while it is not possible for other interactions (strong interaction, electromagnetic interaction, weak interaction). The gravitational interaction is much weaker than other interactions, because of the leak of energy to extra dimensions.

There might be different combinations of fundamental constants in the world of whole dimension. We may be seeing only the world on a four-dimensional space with a suitable combination of fundamental constants. We cannot see extra dimensions, where we cannot exist. We can imagine a train which can move only on a one-dimensional railroad given on a two-dimensional plane.

4.8 The mystery of black holes

Astronomers generally divide black holes into three categories according to their mass: stellar mass, supermassive, and intermediate mass. It is a mystery why stellar mass and supermassive black holes have been observed, but intermediate mass black holes have never been observed [51] as shown in figure 4.12.

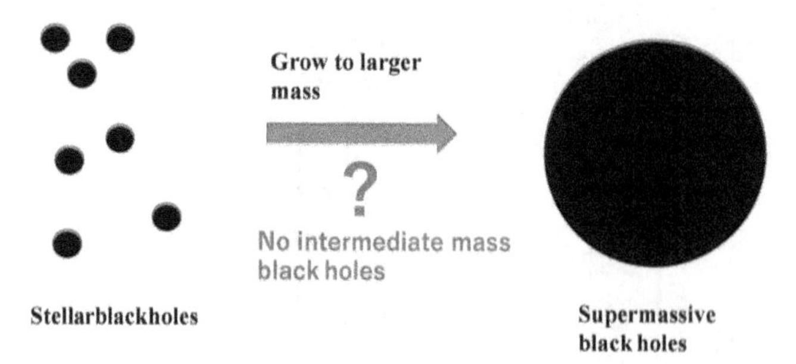

Figure 4.12. The mystery of black holes. Steller black holes are born from the supernova of stars with masses larger than 20 times solar mass. There are also supermassive black holes at the center of almost all large galaxies. But no intermediate mass black holes were observed. Did supermassive black holes came from the increase of mass of stellar black holes by absorbing surrounding material or merging with other stellar black holes?

When a star with more than eight times the solar mass runs out of fuel, its core collapses, rebounds, and explodes as a supernova. If the mass was near the threshold, it creates a city-sized, superdense neutron star. If it had around 20 times the solar mass or more, the star's core collapses into a stellar-mass black hole. However, the mechanism to construct a stellar mass black hole is not known in detail. We don't know the upper limit of the mass of stellar black hole constructed by supernova.

There are supermassive black holes at the center of large galaxies. The mass is in the range of 10^5–10^9 times of the solar mass. It is a mystery how these supermassive black holes came to be. It is reasonable to consider that smaller black holes can grow up by absorbing surrounding objects or merging with other black holes. Gravitational waves emitted by the merging of two black holes are now observed quite often (section 2.10). With this mechanism, it should take a very long period until stellar black holes grow into supermassive ones. However, a supermassive black hole was discovered in a very distant galaxies, which indicates that supermassive black holes were born within 700 million years after the birth of the Universe, which is a serious discrepancy with the previous explanation. Supermassive black holes were constructed within a short period with an unknown mechanism. We don't know if this mechanism still works.

There should also be black holes with intermediate mass (10^2–10^5 times of the solar mass), but they have never been discovered. This result also denies the hypothesis that supermassive black holes came from the growing of solar black holes. If solar black holes grow to become supermassive black holes, there must be a procedure for the intermediate mass. There might be some limit of mass obtained by growing of stellar massive black holes, which has not been clarified yet. To clarify the mechanism to produce black holes, we need much more information about the properties of black holes. For example, we don't know if there is some internal structure with black holes. The validity of the theory of general relativity with ultra-high gravitational potential is not known either.

References

[1] Riess A *et al* 1998 *Astrophys. J.* **116** 1009
[2] Penzias A A and Wilson R W 1965 *Astrophys. J.* **142** 419
[3] Sato K 1981 *Mon. Not. R. Astron. Soc.* **195** 467
[4] Guth A H 1981 *Phys. Rev.* D **23** 347
[5] Perlmutter S, Turner M S and While M 1999 *Phys. Rev. Lett.* **83** 670
[6] Smooth G F *et al* 1992 *Astrophys. J. Lett.* **396** L1
[7] Lee Y-W *et al* 2020 *Astrophys. J.* **903** 22
[8] Kapteyn J C 1922 *Astrophys. J.* **55** 302
[9] Oort J H 1932 *Bull. Astron. Inst. Netherland* **6** 249
[10] Zwicky F 1933 *Helv. Phys. Acta* **6** 110
[11] Rubin V, Thonnard N and Ford W K 1980 *Astrophys. J.* **238** 471
[12] Verlinde E P 2016 *ArXiv* **1611** 02269
[13] Kroupa P 2015 *Can. J. Phys.* **93** 169–202
[14] Aubourg E *et al* 1993 *Nature* **365** 623
[15] Gregory S A and Thomson L A 1978 *Astrophys. J.* **222** 784
[16] van Dokkum P *et al* 2018 *Nature* **555** 629
[17] Garrett K 2010 *Adv. Astron.* **2011** 1–22
[18] Minami Y and Komatsu E 2020 *Phys. Rev. Lett.* **125** 221301
[19] Dirac P A 1937 *Nature* **139** 323
[20] Calmet X and Fritzsch H 2002 *Eur. Phys. J.* D **24** 639
[21] Kozlov M G *et al* 2018 *Rev. Mod. Phys.* **90** 045005
[22] Peik E and Tamm C 2003 *Europhys. Lett.* **61** 181
[23] Bethlem H L *et al* 2008 *Eur. Phys. J. Spec. Top.* **163** 55
[24] Web J K *et al* 1999 *Phys. Rev. Lett.* **82** 884
[25] Murphy M T *et al* 2004 *Astrophysics, Clocks and Fundamental Constants* (Berlin: Springer) p 131
[26] Wilcznska M R *et al* 2020 *Sci. Adv.* **6** eaay9672
[27] Le T D 2019 *Result Phys.* **12** 1035
[28] van Weerdenburg F *et al* 2011 *Phys. Rev. Lett.* **106** 180802
[29] Bagdonaite J *et al* 2013 *Science* **339** 46
[30] Rosenband T *et al* 2008 *Science* **319** 1808
[31] Huntemann N *et al* 2014 *Phys. Rev. Lett.* **113** 210802
[32] Godun R M *et al* 2014 *Phys. Rev. Lett.* **113** 210801
[33] Lange R *et al* 2021 *Phys. Rev. Lett.* **126** 011102
[34] Kajita M *et al* 2014 *Phys. Rev.* A **89** 032509
[35] Kajita M 2017 *Phys. Rev.* A **95** 023418
[36] Hannnecke D 2021 *Quantum Sci. Technol.* **6** 014005
[37] Hees A *et al* 2016 *Phys. Rev. Lett.* **117** 061301
[38] Kobayashi T *et al* 2022 *Phys. Rev. Lett.* **129** 241301
[39] Oswald R *et al* 2022 *Phys. Rev. Lett.* **129** 031302
[40] Barontini C *et al* 2022 *EPJ Quantum Technol.* **9** 1
[41] Abe K *et al* 2013 *Phys. Lett.* B **724** 46
[42] Aprile E *et al* 2022 *Phys. Rev. Lett.* **129** 161805
[43] Aprile E *et al* 2023 *Phys. Rev. Lett.* **131** 041003
[44] Yagi Y *et al* 2023 *IEEE Trans. Appl. Supercond.* **33** 5

[45] Brubaker B M *et al* 2017 *Phys. Rev. Lett.* **118** 061302
[46] Fischbach E *et al* 1986 *Phys. Rev. Lett.* **56** 3
[47] Fischbach E 1992 *Nature* **356** 207
[48] Cartlidge E 2016 *Nature* https://doi.org/10.1038/nature.2016.19957
[49] Feng J L *et al* 2016 *Phys. Rev. Lett.* **117** 071803
[50] Green M B *et al* 1988 *Superspring Theory* (Cambridge: Cambridge University Press)
[51] Maccarone T J *et al* 2007 *Nature* **445** 7124
[52] Kajita M *et al* 2018 *Measuring Time* (Bristol: IOP Publishing)

IOP Publishing

Fundamentals of Modern Physics
Unveiling the mysteries
Masatoshi Kajita

Chapter 5

Conclusion

This book has introduced mysteries which have not yet been solved with modern physics. As experimental technology improved, previous mysteries were solved but this also created new mysteries. This book has mainly introduced mysteries in elemental particles and astrophysics.

The discovery of antiparticles solved the mystery of negative rest energy derived by relativistic quantum mechanics. The relationship between particle and antiparticle is close to the charge conjugation + mirror image (CP-symmetry). But if CP-symmetry holds strictly, the abundance of particles and antiparticles must be exact equal. The violation of CP-symmetry was discovered, and a theory was provided to explain it. But further research is required to describe the matter dominated Universe. It is also a mystery that violation of CP-symmetry has only been observed with the weak interaction, although it can also be violated by the strong interaction from the theoretical derivation.

There was the mystery that the immutable shape of the Universe could not be derived from the theory of general relativity. This problem was solved by the discovery of the expansion of the Universe. The discovery of the cosmological microwave background is evidence of the birth of the Universe with the explosion at one point. But this created another mystery, that of 'dark energy', which accelerates the speed of expansion. Another mystery was discovered with the gravitational interaction being much larger than the prediction from the mass of observed matter. There must be material which has gravitational interaction but no electromagnetic interaction. The identities of dark energy and dark matter have never been identified.

There are also some other puzzles to be solved. There is the subject of the unification of the four kinds of interaction, although the electromagnetic interaction and the weak interaction were unified (electroweak interaction). The grand unification theory to unify the strong interaction with the electroweak interaction might be established by the discovery of proton decay. The unification of the theory

doi:10.1088/978-0-7503-6239-9ch5

of general relativity and quantum mechanics has never been successful. The super string theory was proposed, but there is no experimental evidence.

How can we solve these remaining mysteries? Previously new physics has been developed by observing slight effects after the reduction of measurement uncertainty. Newtonian mechanics was established when the accuracies of clocks was improved significantly by the discovery of the periodicity of the pendulum's swing. The speed of light was measured with high accuracy, which provided the possibility to establish the theory of relativity. Further reduction of measurement uncertainty will contribute to solving the mysteries remaining with modern physics. For example, the variation of fundamental constants might be detected via the precision measurement of atomic or molecular transition frequencies, which might give information about the identity of dark matter. The unequal abundance of the optical isomers of chiral molecules can be explained when the slight difference of vibrational transition frequencies is detected between both isomers.

The expansion of observation range is also required for the development of new physics. Observation of quasars at far distances created new mysteries. Observation of phenomena with ultra-high or low energy also gave us important information. The discovery of particles in higher generations had a large role in establishing a theoretical explanation of the violation of the CP-symmetry. High energy was required for the discovery of particles in higher generations. Laser cooling of atoms and molecules made it possible to observe their wave characteristics, which became significant with ultra-low kinetic energy. Using cold atoms and molecules, the measurement uncertainty of transition frequency was drastically reduced.

As an author, it would be a great pleasure if this book kindles an interest in young researchers to solve the remaining mysteries. After solving a few of these mysteries, a new revolution in physics is expected. There are discrepancies between the descriptions of the neutrino and antineutrino in chapter 3. This is because we don't know the real identity of the neutrino.

IOP Publishing

Fundamentals of Modern Physics
Unveiling the mysteries
Masatoshi Kajita

Appendix A

Boltzmann energy distribution and entropy

In this appendix, fundamentals of statistic mechanics is introduced to derive equation (1.2.11) in the text. We consider energies inside and outside the system E_{in} and E_{out}, where $E_{\text{in}} + E_{\text{out}} = E_{\text{tot}}$ is constant. The numbers of states at inside and outside the system are denoted by Ω_{in} an Ω_{out}. The possibility of the energy distribution is proportional to

$$\Omega_{\text{tot}} = \Omega_{\text{in}}(E_{\text{in}})\Omega_{\text{out}}(E_{\text{out}}). \tag{A.1}$$

Entropy as a parameter of randomness is defined by $S_E = k_B \ln \Omega$. The fundamental of statistical mechanics is the 'increase of entropy'

$$\frac{\mathrm{d}S_{E-\text{tot}}}{\mathrm{d}t} = \frac{\mathrm{d}S_{E-\text{in}}}{\mathrm{d}t} + \frac{\mathrm{d}S_{E-\text{out}}}{\mathrm{d}t} \geqslant 0. \tag{A.2}$$

The thermal equilibrium state is attained when the entropy is maximum.

Equation (A.2) is rewritten as

$$\frac{\mathrm{d}E_{\text{in}}}{\mathrm{d}t}\left[\frac{\mathrm{d}S_{E-\text{in}}}{\mathrm{d}E_{\text{in}}} + \frac{\mathrm{d}S_{E-\text{out}}}{\mathrm{d}E_{\text{in}}}\right] = \frac{\mathrm{d}E_{\text{in}}}{\mathrm{d}t}\left[\frac{\mathrm{d}S_{E-\text{in}}}{\mathrm{d}E_{\text{in}}} - \frac{\mathrm{d}S_{E-\text{out}}}{\mathrm{d}E_{\text{out}}}\right] \geqslant 0. \tag{A.3}$$

Defining the thermal kinetic temperature T by

$$\frac{1}{T} = \frac{\mathrm{d}S_E}{\mathrm{d}E}, \tag{A.4}$$

equation (A.3) is rewritten as

$$\frac{\mathrm{d}E_{\text{in}}}{\mathrm{d}t}\left[\frac{1}{T_{\text{in}}} - \frac{1}{T_{\text{out}}}\right] \geqslant 0 \tag{A.5}$$

which indicates that E_{in} increases (decreases) when $T_{\text{in}} < T_{\text{out}}$ $(T_{\text{in}} > T_{\text{out}})$. With the thermal equilibrium state, $T_{\text{in}} = T_{\text{out}}(=T)$. Equation (A.4) is rewritten as

$$\frac{dS_{E-\text{out}}}{dE_{\text{in}}} = -\frac{dS_{E-\text{out}}}{dE_{\text{out}}} = -\frac{1}{T}$$

$$S_{E-\text{out}} = k_B \ln(\Omega_{\text{out}}) = -\frac{E_{\text{in}}}{T} + \text{const.}$$

$$\Omega_{\text{out}} \propto \exp\left(-\frac{E_{\text{in}}}{k_B T}\right)$$

$$\Omega_{\text{tot}} \propto \Omega_{\text{in}}(E_{\text{in}})\exp\left(-\frac{E_{\text{in}}}{k_B T}\right). \tag{A.6}$$

Equation (1.2.11) is derived with the following procedure. The energy of the electromagnetic wave is only $n_a h\nu$, where ν is the frequency, h is Planck's constant, and n_a is an integer. The average energy of the electromagnetic wave with the frequency of ν is given by

$$P_{BBR} = \Omega(\nu)\sum(n_a h\nu)R_{\text{prob}}(n_a)$$

$$R_{\text{prob}}(n_a) = \frac{\exp\left(-\frac{n_a h\nu}{k_B T}\right)}{Z}$$

$$Z = \sum_n \exp\left(-\frac{n_a h\nu}{k_B T}\right) = \frac{1}{1 - \exp\left(-\frac{h\nu}{k_B T}\right)}. \tag{A.7}$$

$R_{\text{prob}}(n_a)$ denotes the possibility to have the energy of $nh\nu$. Z is a parameter to make $\sum R_{\text{prob}}(n_a) = 1$. $\Omega(\nu)\mathrm{d}\nu$ is the number of states with the frequency of ν. Using

$$(n_a h\nu)\exp\left(-\frac{n_a h\nu}{k_B T}\right) = -\frac{d}{\mathrm{d}\left(\frac{1}{k_B T}\right)}\exp\left(-\frac{n_a h\nu}{k_B T}\right), \tag{A.8}$$

equation (A.7) is rewritten as

$$P_{BBR} = -\Omega(\nu)\frac{1}{Z}\frac{\mathrm{d}Z}{\mathrm{d}\left(\frac{1}{k_B T}\right)} = \Omega(\nu)\frac{h\nu}{\exp\left(\frac{h\nu}{k_B T}\right) - 1}. \tag{A.9}$$

$\Omega(\nu)\mathrm{d}\nu$ is given as follows by the state of polarization 2 and the direction of propagation given by the wavenumber vector $\vec{k} = (k_x, k_y, k_z)$, where $k = \sqrt{k_x^2 + k_y^2 + k_z^2} = \nu/c$,

$$\Omega(\nu)\mathrm{d}\nu = 2 \times \mathrm{d}k_x \mathrm{d}k_y \mathrm{d}k_z = 8\pi k^2 \mathrm{d}k = \frac{8\pi\nu^2 \mathrm{d}\nu}{c^3} \tag{A.10}$$

and equation (1.2.11)

$$P_{BBR} = \frac{8\pi h\nu^3}{c^3} \frac{\mathrm{d}\nu}{\exp\left(\frac{h\nu}{k_B T}\right) - 1} \tag{A.11}$$

is derived.

Section 2.4 compares the entropy of atoms before and after optical pumping. Here we consider the entropy when atoms are distributed to two states A and B. The energy in the B state is higher than A states by ΔE. When the total number of atoms is N, the number of states when n_A atoms are in the A states.

$$\begin{aligned} \Omega_{\text{in}}(n_A) &= \frac{N!}{n_A(N - n_A)} \\ \Omega_{\text{out}}(n_A) &= \exp\left[-\frac{(N - n_A)\Delta E}{k_B T}\right]\Omega_{\text{out}}(N) \\ \Omega_{\text{tot}}(n_A) &= \Omega_{\text{in}}(n_A)\Omega_{\text{out}}(n_A) = \frac{N!}{n_A(N - n_A)}\exp\left[-\frac{(N - n_A)\Delta E}{k_B T}\right]\Omega_{\text{out}}(N) \end{aligned} \tag{A.12}$$

and the entropy is given by

$$S_E = k_B\left[\ln N! - \ln n_A! - \ln(N - n_A)! + \ln[\Omega_{\text{out}}(N)] - \frac{(N - n_A)\Delta E}{k_B T}\right]. \tag{A.13}$$

The entropy is maximum with

$$\begin{aligned} \frac{\mathrm{d}S_E}{\mathrm{d}n_A} &= k_B\left[\ln\frac{N - n_A}{n_A} + \frac{\Delta E}{k_B T}\right] = 0 \\ \frac{N - n_A}{n_A} &= \exp\left[-\frac{\Delta E}{k_B T}\right]. \end{aligned} \tag{A.14}$$

When equation (A.14) is satisfied, the entropy is maximum with the value of

$$\begin{aligned} n_A &= \frac{N}{\exp\left[\frac{\Delta E}{k_B T}\right] + 1} \\ S_E &= k_B \ln\frac{N!}{[n_A!]^2} + k_B \ln \Omega_{\text{out}}(N). \end{aligned} \tag{A.15}$$

After all atoms are localized to the state A or B by optical pumping, the entropy is given by

$$\begin{aligned} &\text{pump to A} \;\; n_A = N, \;\; S_E = k_B \ln \Omega_{\text{out}}(N) \\ &\text{pump to B} \;\; n_A = 0, \;\; S_E = k_B \ln \Omega_{\text{out}}(N) - \frac{N\Delta E}{k_B T}. \end{aligned} \tag{A.16}$$

The entropy of the atom is reduced by optical pumping. However, the laser light with a uniform direction and phase is transformed to the fluorescence light with random direction and phase, and the total entropy is increased.

Appendix B

Energy state of hydrogen atoms

This appendix provides the derivation of equation (1.5.42) in text.

$$ER(r) = \left[-\frac{h^2}{8\pi^2\mu_e}\left(\frac{\partial^2}{\partial r^2} + \frac{2}{r}\frac{\partial}{\partial r}\right) + \frac{h^2 L(L+1)}{8\pi^2\mu_e r^2} - \frac{e^2}{4\pi\varepsilon_0 r}\right] R(r)$$

$$\left[-\frac{\partial^2}{\partial r^2}G(r) - \frac{2}{r}\frac{\partial}{\partial r}G(r) + \frac{L(L+1)}{r^2} - \frac{2}{a_B r}\right] R(r) = \zeta R(r)$$

$a_B = \frac{\varepsilon_0 h^2}{\pi\mu_e e^2}$ (Bohr radius)

$$\zeta = \frac{8\pi^2\mu_e}{h^2}E_e \tag{B.1}$$

We assume the formula

$$R(r) = G(r)\exp(-\alpha r). \tag{B.2}$$

Then equation (B.1) is rewritten as

$$\left[-\frac{\partial^2}{\partial r^2}G(r) + 2\alpha\frac{\partial}{\partial r}G(r) - \frac{2}{r}\left(\frac{\partial}{\partial r}G(r) - \alpha G(r)\right) + \frac{L(L+1)}{r^2}G(r) - \frac{2}{a_B r}G(r) - \alpha^2 G(r)\right] = \zeta G(r). \tag{B.3}$$

Requiring that equation (B.3) as an identity, the following relation is given.

$$-\alpha^2 = \zeta \quad E_e = -\frac{h^2}{8\pi^2\mu_e}\alpha^2$$

$$H_r G(r) = H_{r1}G(r) + H_{r2}G(r) = 0$$

$$H_{r1}G(r) = 2\left[\alpha\frac{\partial}{\partial r}G(r) + \frac{1}{r}\left(\alpha - \frac{1}{a_B}\right)G(r)\right]$$

$$H_{r2}G(r) = -\frac{\partial^2}{\partial r^2}G(r) - \frac{2}{r}\ \frac{\partial}{\partial r}G(r) + \frac{L(L+1)}{r^2}G(r). \tag{B.4}$$

doi:10.1088/978-0-7503-6239-9ch7

We consider with the limit of $r \to \infty$ and $r \to 0$. Note that $H_{r1}G(r)$ and $H_{r2}G(r)$ are functions of r with dimension one and two orders lower than $G(r)$, respectively. With $r \to \infty$, $H_{r2}G(r)$ is negligibly small in comparison with $H_{r1}G(r)$. Therefore, $H_rG(r) = 0$ is obtained by

$$H_rG(r) \approx H_{r1}G(r) = 0$$
$$\frac{\partial}{\partial r}G(r) = \frac{1}{r}\left(\frac{1}{a_b\alpha} - 1\right)G(r),$$
$$\int\frac{dG(r)}{G(r)} = \left(\frac{1}{a_B\alpha} - 1\right)\int\frac{dr}{r},$$
$$\ln G(r) = \left(\frac{1}{a_B\alpha} - 1\right)\ln r + \text{const},$$
$$G(r) \propto r^{n-1} \quad n = \frac{1}{a_B\alpha},$$

$$E_e = -\frac{h^2}{8\pi^2\mu_e}\alpha^2 = -\frac{h^2}{8\pi^2\mu_e}\left(\frac{1}{a_Bn}\right)^2 = -\frac{e^2}{8\pi\varepsilon_0 a_B}\frac{1}{n^2}. \tag{B.5}$$

Equation (1.5.42) corresponds to equation (B.5). The requirement that n is integer (principal quantum number) is derived with the following procedure. With $r \to 0$,

$$H_rG(r) \approx H_{r2}G(r) = 0$$
$$-\frac{\partial^2}{\partial r^2}G(r) - \frac{2}{r}\frac{\partial}{\partial r}G(r) + \frac{L(L+1)}{r^2}G(r) = 0.$$

$$G(r) \propto r^L,\ r^{-(L+1)}. \tag{B.6}$$

Requiring that $G(r)$ does not diverge at $r \to 0$, $G(r) \propto r^L$. As shown in section 1.5 in text, L is required to be integer.

Here we consider the formula of $G_{n,L}(r) = \sum_{q=L}^{n-1} c_q r^q$ with the relation

$$\begin{aligned} c_qH_{r2}r^q &= c_q[-q(q+1) + L(L+1)]r^{q-2},\\ c_qH_{r1}r^q &= \frac{2}{a_Bn}c_q(q-n+1)r^{q-1}. \end{aligned} \tag{B.7}$$

To satisfy $H_rG_{n,L}(r) = 0$, the following relation is required.

$$\begin{aligned} c_qH_{r2}r^q + c_{q-1}H_{r1}r^{q-1} &= \left\{c_q[-q(q+1) + L(L+1)] - \frac{2}{a_Bn}c_{q-1}(n-q)\right\}r^{q-2} = 0\\ c_q[-q(q+1) + L(L+1)] &= \frac{2}{a_Bn}c_{q-1}(n-q), \end{aligned} \tag{B.8}$$

If n is not integer, $c_q \neq 0$ with also negative values of q, while $c_q = 0$ is required with $q < L$ and $q \geqslant n$. Therefore, n must be integer. The solutions of $R_{n,L}(r)$ with (n, L) states are given by

$$R_{1,0}(r) = 2\left(\frac{1}{a_B}\right)^{\frac{3}{2}} \exp\left(-\frac{r}{a_B}\right)$$

$$R_{2,1}(r) = \frac{1}{2\sqrt{6}}\left(\frac{1}{a_B}\right)^{\frac{3}{2}}\left(\frac{r}{a_B}\right) \exp\left(-\frac{r}{2a_B}\right) \tag{B.9}$$

$$R_{2,0}(r) \propto \frac{1}{\sqrt{2}}\left(\frac{1}{a_B}\right)^{\frac{3}{2}}\left(1 - \frac{r}{2a_B}\right) \exp\left(-\frac{r}{2a_B}\right)$$

The motion of the electron in the $L = 0$ state is interpreted as the vibrational motion in the radial direction, because the distribution of the electron wavefunction is not zero at $r = 0$. The electron distribution in the $(n = 1, L = 0)$ state is derived from the uncertainty between the position and momentum in the radial direction. The distribution at small r becomes smaller at higher L state, because of the centrifugal force.

Appendix C

Measurement uncertainty

When performing a physical measurement, there is a question of whether the obtained results are reliable ones. The reliability of measured values can be confirmed by repeating the measurements many times. The measurement results are expected to be distributed over a certain limited range. The uncertainty given by the non-zero distribution area is called the 'statistical uncertainty'. This broadening can be induced by the temporal fluctuation of the event, which can be reduced by stabilizing the circumstances. For the measurement of the atomic or molecular transition uncertainties, the broadening of the spectrum σ_s is determined mainly by the limited measurement time without the phase jump of the wavefunction, caused by the spontaneous emission transition, collision, or the limited interaction time between the atoms (or molecules) and the electromagnetic wave for the probe. The broadening of the spectrum is interpreted as the effect of the uncertainty between time and energy. The central position of the distribution is estimated averaging the samples of measurement results (number of sample N_s). The statistical uncertainty is the uncertainty of the central position, which is roughly estimated by $\sigma_s/\sqrt{N_s}$. For the measurement of the atomic or molecular transition frequency ν_0, the fractional statistical measurement uncertainty is given by

$$\frac{(\Delta\nu)_{\text{stat}}}{\nu_0} = \frac{\sigma_s}{\nu_0}\sqrt{\frac{1}{N_s}} = \frac{\sigma_s}{\nu_0}\sqrt{\frac{\tau_m}{N_a T_m}}, \tag{C.1}$$

where τ_m is a time for the single measurement, T_m is the measurement time, and N_a is the number of atoms or molecules. Equation (C.1) indicates that high transition frequencies with narrow spectrum linewidth is advantageous to reduce the statistic measurement uncertainty. Note that the spontaneous emission transition rate is proportional to the cube of the transition frequencies, therefore, electric dipole allowed transitions in the optical region are not advantageous for precision measurement (σ_s is of the order of several MHz). With electric dipole forbidden transitions in the optical region, σ_s/ν_0 is below 10^{-14} and statistical uncertainty below 10^{-17} can be obtained.

doi:10.1088/978-0-7503-6239-9ch8

For the microwave transition, σ_S is determined by the interaction time between atoms (or molecules) and atoms.

The measurement results might be distributed in an area shifted from the real value because the measurement values can vary according to a given set of circumstances. The real values should be defined under a specified condition, and the measured value is shifted with another circumstance. If we know the dependence of the measurements on the circumstances, we can correct the measured values using the estimated shift. Then the uncertainty of the estimated shift becomes another uncertainty, called the 'systematic uncertainty'. Systematic uncertainty is reduced by monitoring the state of measurement circumstances and correcting the estimated shifts. For the transition frequencies of atoms or molecules, the shifts of the measurements are induced by the electric field (Stark shift) and the magnetic field (Zeeman shift). These shifts are often eliminated by averaging several transition frequencies giving shifts in opposite directions. For example, the linear Zeeman shift in the $(L, J) = (0, 1/2) \rightarrow (2, 5/2)$ transition frequencies of alkali-like ions are eliminated by averaging the $M_J = \pm M'_J \rightarrow \pm M''_J$ transition frequencies (L: quantum number of electron orbital angular momentum, J: quantum number of fine structure, M_J: component of J parallel to the magnetic field). The Stark or Zeeman shifts are eliminated also by estimating electric or magnetic fields by measuring other transition frequencies, with which the Stark or Zeeman shifts are much more significant. Measurements of the transition frequencies are also shifted owing to relativistic effects (quadratic Doppler shift, gravitational red shift shown in section 2.2). The quadratic Doppler shift is below 10^{-17} for the measurement of the transition frequencies of cold atoms or molecules with the kinetic energy below 10^{-3} K.

The total measurement uncertainty is obtained as the square-root of the sum of the squares of each uncertainty (statistical uncertainty, systematic uncertainties cause by different effects).

IOP Publishing

Appendix D

Energy eigenvalue in the harmonic potential

The Hamiltonian of the harmonic potential with the vibrational frequency of ν_{vib} is given by

$$H\Phi = \left[\frac{p^2}{2m} + \frac{m}{2}(2\pi\nu_{\text{vib}})^2 x^2\right]\Phi \tag{D.1}$$

where m is the mass of the particle. For the eigenfunction corresponding to the energy eigenvalue of E_{eig}, equation (D.1) is rewritten as

$$\begin{aligned} &\Phi = \exp\left(\frac{2\pi iE}{h}t\right)\Psi \\ &\left[\frac{p^2}{2m} + \frac{m}{2}(2\pi\nu_{\text{vib}})^2 x^2\right]\Psi = E\Psi \end{aligned} \tag{D.2}$$

Here we consider operators

$$a^{\pm} = \sqrt{\frac{1}{2m}}p \pm i\sqrt{\frac{m}{2}}(2\pi\nu_{\text{vib}})x. \tag{D.3}$$

Then

$$\mathrm{H} = a^{-}a^{+} - \frac{h\nu_{\text{vib}}}{2} = a^{+}a^{-} + \frac{h\nu_{\text{vib}}}{2} \quad a^{-}a^{+} - a^{+}a^{-} = h\nu_{\text{vib}} \tag{D.4}$$

is obtained using equation (1.5.16). When $\Psi(\varepsilon)$ is the eigenfunction with the eigenenergy of ε,

$$
\begin{aligned}
H[a^{\pm}\Psi(\varepsilon)] &= \left[a^{\mp}a^{\pm}a^{\pm} \mp \frac{h\nu_{\text{vib}}}{2}a^{\pm}\right]\Psi(\varepsilon) \\
&= \left[(a^{\pm}a^{\mp} \pm h\nu_{\text{vib}})a^{\pm} \mp \frac{h\nu_{\text{vib}}}{2}a^{\pm}\right]\Psi(\varepsilon) \\
&= a^{\pm}\left(a^{\mp}a^{\pm} \pm \frac{h\nu_{\text{vib}}}{2}\right)\Psi(\varepsilon) \\
&= a^{\pm}(H \pm h\nu_{\text{vib}})\Psi(\varepsilon) \\
&= (\varepsilon \pm h\nu_{\text{vib}})(a^{\pm}\Psi(\varepsilon)),
\end{aligned} \tag{D.5}
$$

and $(a^{\pm}\Psi(\varepsilon)) \propto \Psi(\varepsilon \pm h\nu_{\text{vib}})$ is derived.

We can also give another interpretation of equation (D.5). From the constancy of the total energy ε, the following expression is given:

$$
\sqrt{\frac{1}{2m}}p = \sqrt{\varepsilon}\cos(2\pi\nu_{\text{vib}}t) \quad \sqrt{\frac{m}{2}}(2\pi\nu_{\text{vib}})x = \sqrt{\varepsilon}\sin(2\pi\nu_{\text{vib}}t). \tag{D.6}
$$

Then

$$
\begin{aligned}
a^{\pm} &= \sqrt{\varepsilon}[\cos(2\pi\nu_{\text{vib}}t) \pm i\sin(2\pi\nu_{\text{vib}}t)] = \sqrt{\varepsilon}\exp(\pm 2\pi i\nu_{\text{vib}}t) \\
(a^{\pm}\Phi(\varepsilon)) &\propto \sqrt{\varepsilon}\exp\left[\frac{2\pi i}{h}(\varepsilon \pm h\nu_{\text{vib}})\right] \propto \Phi(\varepsilon \pm h\nu_{\text{vib}}).
\end{aligned} \tag{D.7}
$$

To prohibit the negative value of eigenenergy, minimum energy eigenvalue of $\varepsilon_0 = \frac{h\nu_{\text{vib}}}{2}$ and $a^{-}\Psi(\varepsilon_0) = 0$ are required. $\Psi(\varepsilon_0)$ is obtained by

$$
a^{-}\Psi(\varepsilon_0) = 0
$$

$$
\frac{\partial}{\partial x}\Psi(\varepsilon_0) = -\frac{4\pi^2\mu_a\nu_{\text{vib}}}{h}x\Psi(\varepsilon_0)
$$

$$
\int \frac{1}{\Psi(\varepsilon_0)}\text{d}\Psi(\varepsilon_0) = -\int \frac{4\pi^2\mu_a\nu_{\text{vib}}}{h}x\text{d}x
$$

$$
\Psi(\varepsilon_0) \propto \exp\left(-\frac{2\pi^2\mu_a\nu_{\text{vib}}}{h}x^2\right). \tag{D.8}
$$

The higher energy eigenvalues and functions are given by

$$
\begin{aligned}
\varepsilon_{n_v} &= \left(n_v + \frac{1}{2}\right)h\nu_{\text{vib}} \\
\Psi(\varepsilon_{n_v}) &= (a^{+})^{n_v}\Psi(\varepsilon_0).
\end{aligned} \tag{D.9}
$$

The trap potential is approximately given by the harmonic potential at around $x = 0$. With high energy, the potential energy is not always given by the harmonic potential and equation (D.9) is not accurate.

IOP Publishing

Fundamentals of Modern Physics
Unveiling the mysteries
Masatoshi Kajita

Appendix E

Transition between two or three states

In this appendix, we discuss the transition between two states a and b induced by the AC electric field of light in one direction $E = E_0 \cos(2\pi\nu t)$. The Hamiltonian is given by

$$H = H_0 + H'$$

$$H_0\Phi_{a,\,b} = E_{a,\,b}\Phi_{a,\,b}$$

$$\Phi_{a,\,b} = \Psi_{a,\,b}\exp\left(\frac{2\pi i}{h}E_{a,\,b}t\right)$$

$E_{a,b}$: Energy eigenvalues at a and b state

$$H' = \check{d}E_0\cos(2\pi\nu t) = \check{d}E_0\frac{[\exp(2\pi i\nu t) + \exp(-2\pi i\nu t)]}{2}$$

$$\check{d}\text{: electric dipole moment operator.} \tag{E.1}$$

The temporal change of the wavefunction $\Phi = a\Phi_a + b\Phi_b$ is given by

$$H\Phi = \frac{h}{2\pi i}\frac{\partial a}{\partial t}\Phi_a + \frac{h}{2\pi i}\frac{\partial b}{\partial t}\Phi_b + a\frac{h}{2\pi i}\frac{\partial\Phi_a}{\partial t} + b\frac{h}{2\pi i}\frac{\partial\Phi_b}{\partial t}$$

$$= aH_0\Phi_a + bH_0\Phi_b + aH'\Phi_a + bH'\Phi_b$$

$$\frac{h}{2\pi i}\frac{\partial a}{\partial t}\Phi_a + \frac{h}{2\pi i}\frac{\partial b}{\partial t}\Phi_b = aH'\Phi_a + bH'\Phi_b. \tag{E.2}$$

Here we assume

$$d_{aa} = \iiint \Phi_a\tilde{d}\Phi_a dV = 0$$

$$d_{bb} = \iiint \Phi_b \tilde{d} \Phi_b dV = 0. \tag{E.3}$$

Considering the product of $\Phi^*_{a,b}$ with both sides of equation (E.2):

$$\frac{h}{2\pi i}\frac{\partial a}{\partial t} = bH'_{ab} \quad H'_{ab} = d_{ab}E_0\frac{[\exp(2\pi i(\nu_0 + \nu)t) + \exp(2\pi i(\nu_0 - \nu)t)]}{2}$$

$$\frac{h}{2\pi i}\frac{\partial b}{\partial t} = aH'_{ba} \quad H'_{ba} = d_{ba}E_0\frac{[\exp(-2\pi i(\nu_0 + \nu)t) + \exp(-2\pi i(\nu_0 - \nu)t)]}{2}$$

$$d_{ab} = \iiint \Psi^*_a \tilde{d} \Psi_b \mathrm{d}V$$

$$\nu_0 = \frac{E_b - E_a}{h}. \tag{E.4}$$

Equation (E.4) is simplified by ignoring the fast vibrational effect ($\propto \exp[\pm 2\pi i(\nu_0 + \nu)t]$), called the rotational wave approximation. Then we have

$$\frac{\mathrm{d}^2 b}{\mathrm{d}t^2} = \frac{2\pi i}{\hbar}\left[-2\pi i(\nu_0 - \nu)aH'_{ba} + H'_{ba}\frac{\mathrm{d}a}{at}\right] = -2\pi i(\nu_0 - \nu)\frac{\mathrm{d}b}{\mathrm{d}t} - \frac{4\pi^2 \mid H'_{ab}\mid^2}{h^2}b$$

taking $\Delta_f = \nu_0 - \nu$

$$\Omega_R = \frac{d_{ab}E_0}{h} \text{ (called Rabi frequency)}$$

$$\frac{\mathrm{d}^2 b}{\mathrm{d}t^2} + 2\pi i\Delta_f\frac{\mathrm{d}b}{\mathrm{d}t} + (2\pi\Omega_R)^2 b. \tag{E.5}$$

As the general solution of b,

$$b = \mathrm{e}^{-i\pi\Delta_f t}\left[A\sin\left(\pi\sqrt{\Delta_f^2 + \Omega_R^2}\,t\right) + B\cos\left(\pi\sqrt{\Delta_f^2 + \Omega_R^2}\,t\right)\right]. \tag{E.6}$$

Assuming $b = 0$ with $t = 0$, $B = 0$, and the formula of a using A is given by

$$\begin{aligned} a &= \frac{2}{h\Omega_R \mathrm{e}^{-2\pi i\Delta_f t}}\frac{h}{2\pi i}\frac{\partial b}{\partial t} \\ &= \frac{1}{\pi\Omega_R i\mathrm{e}^{-\pi i\Delta_f t}}A\left[-i\pi\Delta_f \sin\left(\pi\sqrt{\Delta_f^2 + \Omega_R^2}\,t\right) + \pi\sqrt{\Delta_f^2 + \Omega_R^2}\cos\left(\pi\sqrt{\Delta_f^2 + \Omega_R^2}\,t\right)\right]. \end{aligned} \tag{E.7}$$

A is obtained from the condition of $\mid a \mid^2 = 1$ with $t = 0$, and the population in the b state is given by

$$\mid b \mid^2 = \frac{\Omega_R^2}{\Delta_f^2 + \Omega_R^2}\left(\sin\left(\pi\sqrt{\Delta_f^2 + \Omega_R^2}\,t\right)\right)^2, \tag{E.8}$$

which is called the 'Rabi oscillation'. When $\Delta_f = 0$, equation (E.8) is rewritten as

$$a = \cos(\pi\Omega_R t)$$

$$b = \sin(\pi\Omega_R t), \tag{E.9}$$

and

$$2\pi\Omega_R t = \frac{\pi}{2} \to a = b = \frac{1}{\sqrt{2}} \quad \frac{\pi}{2}\text{-transition}$$

$$2\pi\Omega_R t = \pi \to a = 0 \quad b = 1 \quad \pi\text{-transition}$$

$$2\pi\Omega_R t = 2\pi \to a = -1\ b = 0 \quad 2\pi\text{-transition} \tag{E.10}$$

We will now examine the case of the three states a_1, a_2, and b. The energy difference between the $a_{1,2}$–b state is $hn_{01,2}$. When light with a frequency of $\nu_1(\nu_2)$ close to $\nu_{01}(\nu_{02})$ is considered, the a_1–b (a_2–b) transition is induced. What happens when both frequency components are irradiated simultaneously? Taking $\Phi = a_1\Phi_{a1} + a_2\Phi_{a2} + b\Phi_b$, equation (E.2) can be rewritten as

$$\frac{\partial b}{\partial t} = 2\pi i\left[\Omega_{R1}\exp\left(-2\pi i\Delta_{f1}t + i\eta_1\right)a_1 + \Omega_{R2}\exp\left(-2\pi i\Delta_{f2}t + i\eta_2\right)a_2\right]$$

$$\Delta_{f1,2} = \nu_{01,2} - \nu_{1,2} \tag{E.11}$$

$\Omega_{R1,2}$: Rabi frequency (defined in equation (E.5)) between the $a_{1,2} - b$ states.

With random $h_{1,2}$, both transitions are induced independently. However, when

$$\Omega_{R1}a_1 = \Omega_{R2}a_2 \tag{E.12}$$

$$\Delta_{f1} = \Delta_{f2} \tag{E.13}$$

$$\eta_1 - \eta_2 = \pi \tag{E.14}$$

are satisfied, $\frac{db}{dt} = 0$, and both transitions are suppressed. This phenomenon, called 'electric induced transparency (EIT)', is then realized. Comparing the $a_1 \to b$ and the $a_2 \to b$ transition rates, the population ratio between the a_1 and a_2 states converges to equation (E.12). When equation (E.13) is also satisfied, both transitions are suppressed after equation (E.14) is coincidentally satisfied after repeating the laser induced excitation and the spontaneous emission deexcitation with a random phase jump. After the transitions are suppressed, the EIT state is maintained.

As shown in equation (E.12), the population ratio between a_1 and a_2 can be controlled by the intensity ratio of both transition frequency components. By reducing Ω_{R1} adiabatically, the state population can be localized to the a_1 state. This procedure is called 'stimulated Raman adiabatic passage (STIRAP)'.

IOP Publishing

Fundamentals of Modern Physics
Unveiling the mysteries
Masatoshi Kajita

Appendix F

Expansion of a travelling wave with spherical functions

The atomic or molecular wavefunction in a free space is obtained with the Schrödinger equation

$$E\Psi = -\frac{h^2}{8\pi^2 m}\left(\frac{\partial^2}{\partial x^2} + \frac{\partial^2}{\partial y^2} + \frac{\partial^2}{\partial z^2}\right)\Psi. \tag{F.1}$$

Using polar coordinates equation (F.1) is rewritten as (equation (1.5.40))

$$E\Psi = \left[-\frac{h^2}{8\pi^2 m}\left(\frac{\partial^2}{\partial r^2} + \frac{2}{r}\frac{\partial}{\partial r}\right) + \frac{\widetilde{L^2}}{2mr^2}\right]\Psi. \tag{F.2}$$

Here we consider the solution of the wavefunction with a given angular momentum,

$$\Psi(r, \theta, \phi) = R(r)\, Y_L^M(\theta)\Theta_M(\varphi)$$

$$\check{L}^2 Y_L^M(\theta)\Theta_M(\varphi) = \left(\frac{h}{2\pi}\right)^2 L(L+1)\, Y_L^M(\theta)\Theta_M(\varphi) \tag{F.3}$$

$R(r)$ is obtained solving

$$\left[\frac{\partial^2}{\partial r^2} + \frac{2}{r}\frac{\partial}{\partial r} - \frac{L(L+1)}{r^2}\right]R_L(r) = -(2\pi k)^2 R_L(r)$$

$$k = \frac{\sqrt{2mE}}{h}. \tag{F.4}$$

Equation (E.4) is simplified as follows using a function $\chi_L(r) = rR_L(L)$:

$$\left[\frac{\partial^2}{\partial r^2} - \frac{L(L+1)}{r^2}\right]\chi_L(r) = -(2\pi k)^2\chi_L(r). \tag{F.5}$$

With $r \to \infty$, the term of the centrifugal potential is ignored and

$$\frac{\partial^2}{\partial r^2}\chi_L(r) = -(2\pi k)^2\chi_L(r)$$

$$\chi_L(r) \propto \sin[2\pi kr + \delta], \cos[2\pi kr + \delta]$$

$$R_L(r) \propto \frac{\sin[2\pi kr + \delta]}{r}, \frac{\cos[2\pi kr + \delta]}{r}. \tag{F.6}$$

$$r \to 0, \text{ With}$$

$$\left[\frac{\partial^2}{\partial r^2} - \frac{L(L+1)}{r^2}\right]\chi_L(r) = 0$$

$$\chi_L(r) \propto r^{L+1}, r^{-L}$$

$$R_L(r) \propto r^L, r^{-(L+1)}. \tag{F.7}$$

The general solution of equation (F.4) is given as follows:

$$R_L(r) \propto j_L[2\pi kr] \quad j_L\text{: spherical Bessel function}$$
$$r \to 0 \ (2\pi kr)^L,$$
$$r \to \infty \ \frac{\sin\left[2\pi kr - \frac{L\pi}{2}\right]}{2\pi kr} \tag{F.8}$$

or

$$R_L(r) \propto n_L[2\pi kr] \quad n_L\text{: spherical Neumann function}$$
$$r \to 0 \ (2\pi kr)^{-(L+1)},$$
$$r \to \infty \ \frac{\cos\left[2\pi kr - \frac{L\pi}{2}\right]}{2\pi kr}. \tag{F.9}$$

General formula of $R_L(r)$ in free space is given by

$$R_L(r) = \frac{1}{\sqrt{1+\beta_L^2}}(2\pi k)^{\frac{3}{2}}\{j_L[2\pi kr] + \beta_L n[2\pi kr]\}. \tag{F.10}$$

When the space is free from any potential including $r = 0$, $\beta_L = 0$ is required to avoid divergence at $r \to 0$. For $L = 0$, an accurate solution

$$R_0(r) \propto (2\pi k)^{\frac{3}{2}} \frac{\sin[2\pi kr]}{2\pi kr} \tag{F.11}$$

is obtained. For $R_L(r)$ with $L \neq 0$,

$$R_L(r) \propto r^L \quad r \ll \frac{1}{k}, \tag{F.12}$$

which shows that the distribution of the wavefunction at small r becomes smaller as the angular momentum increases because of the centrifugal force.

All wave functions in free space are given by

$$\Psi = \sum p_L R_L(r) Y_L^{M_L}(\theta)\Theta_{M_L}(\varphi) \quad p_L\text{: arbitral coefficient.} \tag{F.13}$$

The plane wave propagating in the z-direction is given by

$$e^{ikz} = \sum i^L 2\sqrt{\pi(2L+1)} \frac{R_L(r)}{2\pi k} Y_L^0(\theta). \tag{F.14}$$

Here, $M_L = 0$ is required since the wavefunction should not depend on φ. A travelling wave in one direction must be the coupling of wavefunctions with different angular momentum, because the direction is defined and the uncertainty principle between angular momentum and direction prohibits the deterministic angular momentum.

Appendix G

Optoelectrical Sisyphus cooling

Optoelectrical Sisyphus cooling is a useful method to reduce the kinetic energy of polar molecules. This method is applicable for multiatomic molecules, with which laser cooling is difficult. Molecules in the electric low field seeking states (positive Stark energy shift with electric field) move between places with high and low electric fields as shown in figure G.1. The Stark energy shift depends on the molecular rotational state, and we consider two states with the strong and weak Stark energy shifts: $|S\rangle$ and $|W\rangle$. At the low electric field region, all molecules are localized to the $|S\rangle$ state by optical pumping via the transition to the vibrational excited state and the

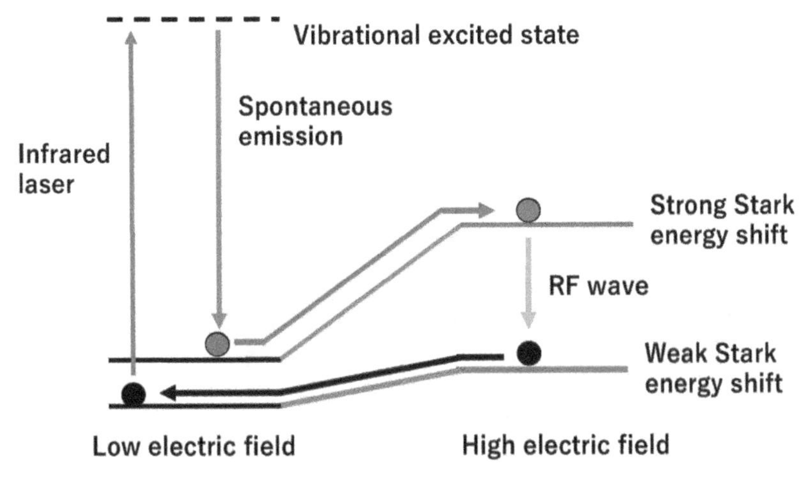

Figure G.1. The procedure of the optoelectrical Sisyphus cooling: molecular cooling by the motion between the high and low electric field region. At low electric field region, molecules are pumped to the state with a strong Stark energy shift and move to the high electric field region getting a strong deceleration. In the high electric field region, molecules are transformed to the state with a weak Stark energy shift by an RF wave. Molecules are accelerated when they move to the low electric field region. The deceleration moving to the region with high electric field is much stronger than the acceleration moving to the region with low electric field and the molecular kinetic energy decreases. Reproduced from [1].

spontaneous emission transition (section 2.4). When molecules in the $|S\rangle$ state moves to the high electric field region, there is an increase of the Stark potential energy and decelerated. Then molecules are transformed to the $|W\rangle$ state using an RF-wave. Molecules in the $|W\rangle$ moves to the low electric field region with an acceleration. With this cycle, molecules are decelerated, because the deceleration in the $|S\rangle$ state is larger than the acceleration in the $|W\rangle$ state.

The kinetic energy of 420 μK and the single rotational state with the purity of 80% was obtained with H_2CO molecule. It is not realistic to cool molecules from the room temperature with this method, and this experiment was performed using slow molecules filtered by the bent waveguide giving two-dimensional trap force by an inhomogeneous electric field.

Reference

[1] Kajita M 2020 *Cold Atoms and Molecules* (Bristol: IOP Publishing)

Printed in the USA
CPSIA information can be obtained
at www.ICGtesting.com
JSHW060738120224
56719JS00007B/43